ASVAB

Study Guide

Ace the Armed Services Vocational Aptitude Battery Exam on Your First Try with No Effort | Test Questions, Answer Keys & Tips to Score a 98% Pass Rate

Anthony J. Kohler

Table of Contents

SECTION 1
GETTING STARTED
WITH THE ASVAB

Chapter 1

The ASVAB Exam

The Armed Services Vocational Aptitude Battery is a standard test that the United States Military uses to evaluate an individual's aptitudes and abilities for different military career paths. This chapter provides an overview of ASVAB, including information about the different versions, subtests, what to expect during the exam day, deciphering ASVAB scores, retaking exams, and how scores affect military training programs and jobs assignments. We will also discuss the different ways each branch of military calculates line scores.

There are two versions of the ASVAB test, the CAT (Computerized Adaptive Test), and the P&P (Paper and Pencil Test). The two versions differ in the administration methods and how questions are presented to test takers. Below is a description of the two versions of the ASVAB test:

CAT ASVAB (Computerized adaptive Test)

This is the most commonly used version. The exam is administered using a computer, and uses an adaptive testing method. The difficulty of questions is adapted to the performance of the test-taker. As the test proceeds, the difficulty of subsequent questions is adjusted according to the accuracy of the previous answers. The next question could be more difficult if the test-taker answered a question correctly. If they responded incorrectly, it may result in an easier question. The adaptive process is repeated throughout the test to tailor the questions according to each individual's level of ability. The adaptive nature allows the CAT-ASVAB to provide a more accurate and efficient assessment of a test-taker's abilities and aptitudes.

P&P ASVAB (Paper and Pencil Test)

As the name implies, this test is administered in a paper and pencil format. The test is based on a set of fixed questions, which are the same across all test takers. The P&P ASVAB is not like the CAT ASVAB in that it does not adjust the difficulty level based upon individual responses. All test takers receive the same questions, in the same order. This version is used when computerized testing equipment is not available in certain situations, like remote locations or specific circumstances.

The ASVAB is designed to evaluate an individual's abilities and aptitudes in different areas regardless of how it is administered. The test content and subtests are the same whether you take the CAT or P&P ASVAB. The specific format and method of administration may vary depending on testing centres and availability.

The ASVAB test consists of nine subtests that measure various aptitudes and skills relevant to different military career fields. Take a look at the subtests in more detail:

General Science (GS). The General Science subtest measures the test taker's understanding of physical and biological science. The test includes questions on biology, chemistry and physics as well as Earth science. The test-takers will be expected to demonstrate their knowledge of scientific principles, and to apply their scientific knowledge in real-world situations.

Arithmetic Reasoning: This subtest assesses a person's ability to solve mathematical problems and apply reasoning. The test includes questions that include arithmetic, word problems and mathematical concepts like percentages, ratios and proportions. The test-takers' ability to analyze and interpret mathematical information is assessed.

Word Knowledge (WK). The Word Knowledge subtest assesses a test taker's understanding of words and vocabulary. The test includes questions to assess a test-taker's ability in determining word definitions, identifying synonyms and antonyms, as well as comprehending word usages in different contexts. This subtest measures language skills as well as the ability to understand and interpret written material accurately.

Paragraph Comprehension: This subtest assesses a test taker's ability to read and comprehend paragraphs. The test presents paragraphs or passages, followed by questions to assess the ability of the test-taker to understand and interpret information. The test-taker's ability to understand main ideas, make inferences, connect logically, and analyze written material is evaluated.

Mathematics Knowledge (MK). The Mathematics Knowledge subtest measures the test taker's grasp of mathematical concepts and their applications. The test covers topics like algebra, geometry and measurement. It also includes statistics, probability and probabilities. The test-takers' ability to solve problems, apply formulas, and show a solid understanding of various mathematical principles is assessed.

Electronics Information (EI). The Electronics Information test assesses the knowledge of electrical devices and systems. The test includes questions on basic electrical circuits and components, symbols, principles, etc. The test-taker's ability to grasp electrical concepts, analyze circuit diagrams and identify electrical components is evaluated.

Auto and Shop Information: This subtest measures the test taker's understanding of basic metalworking and woodworking principles, as well as their knowledge about automotive maintenance and repairs. The test includes questions about automotive systems, shop practices, mechanical components and tools. The test assesses the knowledge of automotive terms, repair techniques and general shop safety.

Mechanical Comprehension: This subtest measures a person's knowledge of mechanical devices and principles. The test includes questions to assess your knowledge of mechanical systems such as pulleys and gears. The test-taker's ability to understand mechanical concepts, read diagrams and solve problems relating to mechanical principles is evaluated.

Assembling Objects: This subtest assesses a test taker's spatial reasoning and mental manipulation skills. The test presents illustrations or diagrams of objects to be assembled or rearranged. The test-taker's ability to visualize spatial relationships and understand how parts fit together is assessed.

The nine subtests provide a comprehensive assessment to determine an individual's abilities and skills. This helps identify strengths and areas of expertise in the military. Understanding the format and content of each subtest will help you prepare effectively and achieve a competitive ASVAB score.

Test-takers experience a carefully constructed environment to ensure fairness and integrity of the ASVAB exam. When they arrive at their testing site, they feel a sense anticipation and are ready for the challenges that lie ahead.

The first thing that test takers encounter is a set of guidelines and instructions. These instructions are a guide to help you navigate the exam and understand the expectations. These guidelines are important for test takers to read carefully as they contain information about the format of the exam, the time limit, the rules and the specific procedures that must be followed. Test-takers who are familiar with the instructions can be confident and well-prepared for the test.

ASVAB exams are usually administered in a proctored setting, which ensures that everyone has a standard and equal testing experience. The ASVAB exam is usually administered at a military entry processing station (MEPS), or a designated test centre. Proctors are professionals who monitor the test takers and ensure that testing conditions are optimal. They help create an environment of concentration and focus, so that all test takers have the same opportunity to demonstrate their abilities.

The ASVAB test is heavily reliant on time management. The ASVAB exam has a time limit for each subtest. It is important that test takers complete all questions within this time. The time limit is a pressure factor that forces test takers to balance speed and accuracy. The time limit allows the test taker to pace themselves, and allot enough time for each subtest. This will allow them to complete the entire test within the specified time.

The ASVAB exam can vary in length depending on which version is administered. The CAT (Computerized Adaptive Test), and P&P (Paper and Pencil Test), have different durations. The CAT-ASVAB adapts to the ability of the test-taker and can result in a variable time frame, while the P&P ASVAB may follow a set of fixed questions and have a predetermined duration. Test-takers should be aware of which version they're taking and its duration to help them manage their time.

The ASVAB exam is the culmination of your preparation, focus, and dedication. Test-takers are given instructions and guidelines before entering a proctored setting where they complete the test within time limits. It is important to be aware of all these factors, and embrace the challenges that are presented on this day. This will increase the chances of success and open doors for a rewarding military career.

The ASVAB score is an important part of the exam and provides valuable information on an individual's qualifications and abilities for military service. The scores are presented in different formats including standard scores and percentile scores. Each format serves a particular purpose when determining a test taker's aptitudes, as well as their eligibility for enlistment or job placement.

The Armed Forces Qualification Test score (AFQT), which is derived from ASVAB, is one of most important scores. The score is based on four subtests, including Arithmetic Reasoning (Arithmetic Reasoning), Word Knowledge (Paragraph Comprehension), and Mathematics Knowledge. The AFQT is a standard measure that determines a test taker's ability to enlist in the military. The AFQT is used to evaluate a candidate's academic and cognitive abilities. A higher score on the AFQT indicates that a candidate will be more likely to meet the minimum requirements for enlistment. The AFQT is a gateway that determines whether an individual has the necessary educational and cognitive requirements to join the military.

The ASVAB also reports composite scores in addition to the AFQT. The composite scores are a breakdown on a test taker's performance for specific subtests or areas of the exam. These composite scores are intended to give a detailed evaluation of the candidate's strengths, weaknesses and abilities. They can then be used to help match a candidate with a military career that matches their interests and abilities. The composite scores are calculated from the combined scores of various subtests that relate to a specific skill set or knowledge area. Composite scores can include scores in Mechanical Maintenance, General Science and Electronics, among other things. These scores allow military recruiters identify individuals with the skills and aptitudes required for certain military occupational specialties or job roles.

ASVAB scores also include standard scores and percentiles. The standard scores are also called scaled scores and they allow a comparison of a test taker's performance to a reference group. These scores are standardized to a mean score of 50 with a standard deviation value of 10. This allows for an accurate comparison of test results across subtests. Percentile scores indicate the percentages of test takers who scored lower than a specific individual. A percentile score of 70%, for example, means that the test taker scored higher than the reference group.

ASVAB test scores are a comprehensive evaluation of the abilities and qualifications a person has for military service. The AFQT is used to determine eligibility for enlistment, and it also serves as a major criterion in job placement. Composite scores provide a breakdown of the performance in certain areas and help match candidates to suitable military career fields. Standard scores and percentiles provide standard comparisons and show where the test taker is compared to others. These scores inform both military recruiters as well as candidates of an individual's abilities and can help them choose the right military occupation specialties.

Test-takers who are not satisfied with their ASVAB score can retake it. There are restrictions and guidelines regarding the number of times one can take the test. It is important to check with the testing centre or a recruiter for the most accurate information.

To ensure fairness and integrity, the military and testing centres have set up rules. The guidelines usually dictate that a test taker must wait a certain amount of time before they can retake ASVAB. This waiting period allows individuals to develop their skills and knowledge in between tests, instead of relying on immediate retesting.

You should seek advice from the recruiter or testing centre about the policies that are in place regarding retaking the ASVAB. The recruiter or testing centre will be able to provide you with accurate information regarding the ASVAB retake policy, the waiting period and any restrictions on retakes. They can also explain the steps involved in scheduling a test retake. The test takers will be well informed and able to make the right decisions about their retake strategies.

If a test taker is not satisfied with his or her initial ASVAB score, they can retake it. There are guidelines that may differ on when one can retake a test. It is important to speak with a testing centre or a recruiter to get accurate information.

ASVAB scores are crucial in determining your military career by determining what training programs and jobs you're eligible for. These scores are used by military recruiters to make decisions and match people with jobs that fit their skills and abilities. It is important to understand how ASVAB results impact your career options.

The minimum ASVAB scores for different military occupations can vary across branches. Each branch has its own set of standards that are based on the cognitive abilities and academic skills required for a successful performance in a particular job role. These requirements are designed to ensure that people entering certain occupations have the knowledge and skills necessary to perform their duties.

The Armed Forces Qualification Test score (AFQT), which is derived from the four ASVAB subtests of (Arithmetic Reasoning), Word Knowledge, Paragraph Comprehension, and Mathematics Knowledge, is very important. The AFQT is a standard measure of academic and cognitive ability and determines your eligibility to enlist in the military. The higher your AFQT scores, the more likely you are to meet the minimum requirements for enlistment.

Specific military occupational specialties may also have criteria other than ASVAB scores. These criteria may include physical fitness requirements, security clearances or medical requirements. These additional criteria are essential to qualify for certain military jobs.

Your ASVAB score, and especially your AFQT, will directly affect the types of military occupations you can choose from. Higher scores can provide you with a wider range of career options and allow you to explore different military careers. Lower scores can limit your options for job assignments. Understanding the score thresholds and requirements for the MOSs that you are interested will allow you to better align your studies and improve your chances of qualifying.

ASVAB scores have a significant impact on the training programs you can attend and the job positions that are available to you. The AFQT is the primary factor used to determine eligibility for enlistment. Specific job assignments may have additional criteria in addition to ASVAB scores. Knowing the requirements for your chosen military occupation specialties and the other criteria will help you make informed decisions and pursue the opportunities that match your skills and interests.

Each branch of the military (Army, Navy, Air Force, Marines, Coast Guard) uses a unique set of ASVAB lines scores to evaluate qualification for different job specialties. The line scores are calculated by giving different weights to the ASVAB subtest results. This allows each branch to evaluate an individual's suitability and aptitude for a particular military occupation. Understanding line scores, and how they relate your desired branch or career field, is essential to determining your eligibility for certain military jobs.

Each branch sets unique line score requirements by assigning different weights for ASVAB subtest results. This is done to reflect the various Military Occupational Specialties (MOSs) within each organization. These line score requirements were designed to identify people who have the aptitudes and skills required to perform well in a specific job role. A technical MOS, for example, may give greater weight to subtests related to mechanical understanding and electronics knowledge while a language MOS might prioritize scores on word knowledge and paraphrase comprehension.

It is important to understand how line scores will affect your eligibility. Official publications and other resources provide detailed information about the line score requirements of various MOSs. You can learn more about the subtests that are most important for your career by studying these requirements.

Understanding the line score requirements will help you focus your preparation on the subtests that are relevant to the career field of your choice. You can customize your study plan by identifying areas where you need to improve your knowledge and skills. This will increase your chances to meet the line score requirement for your chosen MOS.

Line score requirements can vary between branches and even over time, depending on the demands and needs of the military. It is important to stay informed about any updates or changes in line score requirements. This will ensure that you are making the best decisions possible for your military career.

Line scores, derived from ASVAB scores on subtests, are critical in assessing the qualifications of different job specialties for each branch. Each branch has its own weightings for subtest scores in order to determine the line score requirements that are specific to them. You can better prepare for your desired career by researching the line score requirements for your branch. You can improve your line score requirements by improving your subtest scores.

This chapter provides a comprehensive review of the ASVAB test, including important topics like understanding the ASVAB version, mapping the subtests and knowing what to expect during the exam. It also covers how scores are used to determine military training programs, job assignments and retaking the ASVAB. This chapter is designed to provide readers with the knowledge and insight they need to successfully navigate the ASVAB test and make informed career decisions.

Chapter 2

How To Pass The Asvab Exam

The ASVAB test is a vital part of determining your career options and eligibility in the United States Military. It is crucial to overcome test anxiety and develop effective strategies in order to excel on this exam. This chapter will provide you with useful tips and techniques to help you do well on the ASVAB.

Many test takers experience test anxiety, which can negatively impact their performance. Using different strategies to manage stress and perform well can help you overcome test anxiety. A thorough preparation is one effective strategy. You can reduce your anxiety and boost your confidence by familiarizing yourself with both the content and the format of the ASVAB test. Understand the format of the test, the questions and the content areas covered. You will feel more confident and comfortable when you are able to approach the test with confidence.

Test anxiety can be managed by using relaxation techniques in conjunction with preparation. Relaxation techniques like progressive muscle relaxation, breathing exercises that deepen your lungs and visualization can help calm nerves. They will also help you relax during and before an exam. These techniques can be practiced in the weeks leading up to the exam. This will help you develop calmness and focus.

Positive self-talk can also be a powerful tool to overcome test anxiety. Positive affirmations can replace negative thoughts and self-doubt. Remember that you put in effort to prepare and study for the exam and that you can succeed. A positive attitude can boost your self-confidence and reduce anxiety.

Time management is essential to a successful ASVAB performance. You should plan how you'll manage your time, and allocate enough time to each section. Don't rush through questions or spend excessive time on one. If you encounter a particularly difficult question, mark it and move on to the next. Manage your time effectively to make sure you have time to answer all the questions and review your answers before you submit your exam.

A good score is achieved by strategizing your approach to the test. To avoid making mistakes, read the instructions carefully for each section. You can maximize your score by avoiding difficult questions and focusing only on those you can answer accurately and quickly. Review your answers to the test before you submit it. This will help you catch any mistakes or missed questions.

The ASVAB test consists primarily of multiple-choice answers. It is essential to have a strategy in place before you begin answering these questions. Before you look at the answers, read the question completely and ensure that you understand it. By eliminating obvious incorrect answers, you can increase your chances of choosing the right one. You can identify the correct response by looking for context clues in the passage or question. Be sure to pay attention to important terms like "not", "except" or "always", as they may have a significant impact on the answer.

It is important to be strategic when you do not know the answer. You can make an educated guess using any information or clues that are provided in the question. To increase your chances of getting it right, eliminate options that are obviously incorrect. It is better to guess if you are uncertain than to leave the questions unanswered. Mark the question that is difficult and move on. If you have time, you can return to the question later.

A good score is achieved by studying and practicing the ASVAB test. Understanding the format of the test and the topics that will be covered is essential. Use study materials such as books, online material, practice tests, and guides designed specifically for the ASVAB. Joining study groups and finding partners with whom to discuss concepts is a good idea. Use practice tests to figure out your strengths and evaluate your area weaknesses.

It is essential to create your own study schedule for an effective preparation. Create a schedule of study that is tailored to your preferences and needs. Assess your strengths, weaknesses and realistic goals. Use different study methods to make your sessions more engaging and productive. Divide your content into smaller topics, and focus on the most challenging ones. Review previously studied material regularly to ensure you retain the information.

You can improve your chances of achieving high scores on the ASVAB test by following these tips. Success is achieved by a combination of a consistent effort, effective strategies and a positive attitude.

It is important to answer multiple-choice questions correctly when taking the ASVAB test. This type of question requires a strategic approach in order to increase your odds of choosing the right answer. Using the strategies below, you can answer multiple-choice questions accurately and with confidence.

It is important to first and foremost read the question completely. Before you look at the answers, take the time to read the entire question. You will be able to make an informed decision if you are armed with a thorough understanding of the issue.

Use the elimination process to your advantage. Eliminate any obvious incorrect answers. This will help you narrow down your choices and increase the chances of choosing the right answer. You can eliminate the answers that don't match the information in the question, or those you know are incorrect. This method will increase the efficiency of the decision-making process, and allow you to concentrate on the remaining possible answers.

Context clues in the passage or question can help you identify the correct answer. Attention to details and hints that can guide you in the right direction. The context can provide valuable information to help you make a well-informed choice. You may want to consider any connections or relationships between the different parts of the questions or relevant information in the passage. This could help you find the right answer.

Be aware of any key words that may have a significant impact on the meaning of a question or its options. The words "not", "except" or "always", can completely change the context and meaning of the question. By paying attention to these words, you can ensure that the question is correctly interpreted and that the answer chosen accurately reflects the conditions.

A systematic and strategic approach is required to effectively tackle multiple-choice ASVAB questions. You can improve your accuracy by reading the question thoroughly, eliminating obvious incorrect answers, using context clues and paying attention to key terms. When combined with practice and thorough preparation, these strategies will help you to navigate the multiple-choice question with confidence.

It can be daunting to encounter a new question on the ASVAB test. There are some smart strategies you can use to deal with such situations.

Make an educated guess first. Use any information or clues in the question, even if you are unaware of the answer. You can use the context of the question, familiar concepts or logic to help you find the answer. By eliminating the obvious incorrect options, you can increase your chances of selecting the right answer.

Don't get stuck if a question is particularly difficult or time-consuming. Mark the question, and then move on. You can better manage your time by skipping the question and coming back to it later. You should manage your time well so you can return to the marked questions at the end of the exam.

When faced with difficult questions, managing your mindset becomes crucial. Maintain a calm, positive attitude and remain calm throughout the test. One difficult question is not indicative of your overall performance. Do not let your frustration or anxiety over a single difficult question impact your confidence, and negatively affect your performance in the remainder of the test. Each question should be approached with a new mindset and the best effort.

It is important to devote enough time to study and practice to perform well on the exam. To improve your preparation, consider the following tips:

Understand the exam structure. You should familiarize yourself with all the different content areas and types of questions you'll encounter on the ASVAB. You can then tailor your study plan to suit this knowledge.

Use a variety study resources. They are useful tools. These materials provide extensive content coverage, practice tests, and explanations to help you better understand and prepare for the ASVAB.

Join study groups, or find study partners. It can be very beneficial to work with others who are preparing for ASVAB. Discussions, the sharing of study materials, questions and explanations can help you to gain a deeper understanding.

Practice tests are a great way to improve your skills. Practice tests allow you to gauge your strengths and weaknesses. You can become familiar with ASVAB's format, timing and type of questions. You can adjust your study schedule based on your analysis of your results.

A personalized study plan will help you prepare effectively. When creating your study schedule, consider the following:

Assess your strengths. Determine the areas in which you excel, and those where more attention is needed. This will help you allocate your time to study more effectively and concentrate on the areas where you need improvement.

Set realistic goals. Divide the ASVAB score into manageable, smaller goals. Setting specific goals helps you track your progress and stay motivated.

Create a schedule for your studies. Each day, allocate specific study time slots. Assure a balanced approach and allocate enough time to cover each content area. A structured schedule of study will allow you to maintain consistency and maximize your study sessions.

Utilize different study techniques. Mix reading, taking notes, practicing questions, using flashcards and interactive methods with reading. Using different study methods keeps sessions interesting and improves retention.

Consider the following steps to create a comprehensive strategy for your study:

Divide and conquer. Divide the content into smaller topics, and then allocate time for study. This allows you to concentrate on specific concepts or subjects, which makes your study more manageable.

Prioritize the difficult areas. Spend more time on subjects or topics you find challenging. You can improve your performance by dedicating more time and resources to the areas you find challenging.

Review material you have already studied. Regularly revisit the material you've already studied. It helps you to reinforce your understanding and refresh your memory.

Keep track of your progress. Record your study sessions, test scores and areas for improvement. You can identify patterns and analyze your strengths and weakness by tracking your progress.

To pass the ASVAB test, a consistent effort, effective strategy, and a positivity mindset are essential. You can increase your chances of passing the ASVAB test by using smart tips to help you answer unfamiliar questions. You can reach your goal with a focussed mindset and diligent preparation.

SECTION 2
COMMUNICATION
SKILLS

Chapter 3

Word Knowlegde

Word knowledge is an essential skill that holds significant importance in our lives. It plays a crucial role in education, communication, and professional success. Understanding words and their meanings enables us to comprehend written texts, express ourselves clearly, and engage in meaningful conversations. Whether we are reading a book, writing an email, or participating in a job interview, word knowledge enhances our language proficiency and increases our chances of success in various endeavours. It goes beyond memorizing definitions and involves understanding how words function in different contexts, allowing us to effectively convey our thoughts, comprehend complex information, and excel in academic and professional settings.

The importance of word knowledge is evident in many aspects of life, such as education, communication and professional success. Understanding words and their definitions helps us understand written texts, communicate clearly and have meaningful conversations. Word knowledge is also crucial for exams and assessments that assess vocabulary, like the ASVAB. We can improve our language skills and our chances of succeeding in different endeavors by improving our word-knowledge.

It is important to become familiar with the format of the questions when preparing for a word knowledge test like the ASVAB. In these tests, we are often asked to choose the best synonym or antonym or identify the meaning of the word. This format will help us approach questions like these with more confidence and better understand what's being asked. We can learn a variety of words and phrases by reading and understanding different texts. This exposure broadens our knowledge and helps us prepare for word knowledge questions.

Another valuable way to decode word meanings is by building words from the ground up. If you are unsure of the meaning of a word, it is helpful to break them down into smaller pieces, such as roots, prefixes and suffixes. Understanding these components, and their common associations, can help you make educated guesses when it comes to unfamiliar words. Understanding the meaning of a word can be further enhanced by looking at the context and finding similar words or concepts.

Word knowledge is incomplete without understanding synonyms and antonyms. Words that have opposite meanings are antonyms. When answering word knowledge questions, it is beneficial to have a good understanding of synonyms. Finding synonyms is a great way to help determine the meaning of a new word. You can do this by looking for a word that has a similar meaning. Antonyms can help eliminate incorrect answers or narrow down options. By expanding our vocabulary of synonyms and antonyms, you can improve your ability to understand word meanings. You will also be able to make better choices when taking word knowledge tests.

It is important to not panic when we come across unfamiliar words in exams or assessments. We can use strategies to make educated assumptions. It can be helpful to analyze the surrounding sentences or words for context or clues about the meaning of the word. Consider the root of the word, its prefixes and suffixes as well as related words that we may be familiar with. If we're unsure about the meaning of a word, the process of elimination will help us eliminate the obvious wrong answers. Making an educated guess can be better than not answering a question.

It is a continuous process which requires constant effort and practice. By reading widely across different materials such as books, articles and newspapers, we are exposed to a diverse vocabulary. We also learn new words and how to use them. Expanding our vocabulary is possible by actively using context clues when reading to understand the meaning of words. Keeping a vocabulary diary to record and learn new words and incorporating them in our writing and speaking are also effective ways to do this. Word games such as word puzzles and word association can be a fun way to improve your vocabulary. Apps that provide interactive games and exercises can also be useful in the quest to improve our vocabulary.

Word knowledge is a vital skill that can impact our academic pursuits, communication skills, and professional endeavours. Understanding words helps us read and understand written texts, communicate effectively, engage in meaningful conversation, etc. In order to improve our performance on word knowledge tests we should familiarize ourselves with the format of the questions and practice diverse reading. Word knowledge tests can be improved by using strategies like deciphering the meaning of words from their parts, understanding synonyms and antonyms, or making educated guesses. We can improve our vocabulary by reading, using context clues and maintaining a vocabulary journal, as well as playing word games and utilizing apps that build vocabulary. This will help us to expand our knowledge of words and enhance our language skills, leading to better academic and professional success.

Chapter 4

Paragraph Comprehension

The ability to read and understand paragraphs is essential for many military careers. It is important for military personnel to be able to read and comprehend written material efficiently, as they often have to deal with complex instructions, manuals and operational procedures. Individuals who have a solid grasp on paragraph comprehension are better able to understand and apply information, make informed choices, and carry out their duties effectively. Paragraph comprehension is essential for military readiness.

Eyeing The Physiques of Paragraph Comprehension Subtest

Paragraph comprehension is an important component of tests like the ASVAB. These are designed to assess the aptitude of people who want to join the military. This subtest measures a candidate's capacity to understand and analyze written passages. This subtest is designed to measure a candidate's ability to understand and analyze written passages effectively.

The subtest includes a variety of written passages. These may include informational articles, narratives, technical passages or other types texts that are commonly encountered in military settings. The subtest includes a variety of texts to ensure that candidates can read and understand a wide range of materials they will encounter in their military career.

Candidates must have strong reading comprehension skills to perform well in the subtest on paragraph comprehension. It is important that they are able to understand and read written passages, even if the language used, the technical terms, or subject matter are unfamiliar. The candidates must be able quickly to grasp the main point of a passage, and to identify the supporting details which contribute to its overall meaning. They should also be able to make logical inferences from the information in the text.

Candidates must read extensively and be exposed to many different texts in order to develop the skills necessary for paragraph comprehension. By reading different materials such as novels or newspapers, technical manuals or scientific articles, candidates can become more familiar with writing styles, vocabulary, and subject matter. This exposure helps candidates become more confident in extracting information, understanding complicated ideas, and analyzing the written content.

Candidates should also improve their ability to understand and identify the main idea in a passage. The main idea is the central theme or concept of the text. It serves as the foundation for understanding the supporting details. Candidates can improve their comprehension skills by identifying and recognizing the main idea in a passage.

Candidates should develop strategies to identify and interpret supporting details in addition to the main idea. These details are evidence, examples or explanations which support the main idea in the passage. Candidates can improve their understanding by improving their ability to analyze and recognize these details.

Inferences are another important aspect of the subtest on paragraph comprehension. Inferences are logical conclusions that can be drawn from the passage even if it is not explicitly stated. Candidates should learn to read between lines, use context and their critical thinking skills in order to deduce meaning and implications.

The subtest of paragraph comprehension is an important part of military assessments, such as ASVAB. The paragraph comprehension test evaluates candidates' abilities to read and understand written passages. This subtest can be improved by candidates practicing how to read, understand, and identify different types of texts, as well as identifying main ideas, recognizing supporting details, drawing logical inferences, and identifying main ideas.

Sample Four Types of Comprehension Questions

In general, the four types of questions asked in a paragraph comprehension test are: main idea (or theme), supporting details (or inferences), vocabulary, and inference. Candidates are asked to identify the main concept or theme in a passage. In order to answer questions that require supporting details, you must find specific information which supports the main concept. Candidates are asked to make logical inferences based on information given in the passage. Vocabulary Questions assess your understanding of words and phrases in context. It is important to familiarize yourself with these types of questions and practice their application in order to achieve success on paragraph comprehension assessments.

Do you get my point? Breaking down paragraphs

It is essential to understand paragraphs by developing strategies for deconstructing them and extracting the key information. Read the passage attentively and pay attention to its main idea. Topic sentences are often the most important part of a paragraph. Search for details or examples that support the main point. To solidify your understanding, use mental summarization techniques such as restating the main idea using your own words. Use these strategies to help you better understand and analyze paragraphs.

How to Analyze what you've read: Guessing what the writer truly meant

Understanding the writer's intentions and implied meanings is crucial to paragraph comprehension. The writer can sometimes convey information in an indirect manner or suggest certain ideas, without explicitly stating these. It is important to make logical deductions from the information given. To decipher what the author intended, look for clues within the text such as the tone, the language used, and the context. Making connections and educated guesses will help you discover the deeper implications of the passage.

How to read faster than a turtle

There are some techniques you can use to become a faster reader if you struggle with it. Practice active reading first by engaging the text and highlighting important information. Take notes while you read. Gradually increase the length of your reading sessions to improve your stamina. To reduce the amount of time you spend on unfamiliar words, expand your vocabulary. Use a pointer to help you focus and guide your reading. You can improve your reading speed by practicing consistently.

Reading and Gleaning Test Tips

It is crucial to manage your time efficiently when attempting paragraph comprehension tests. To get an overview of the passage's content and structure, scan it quickly. Underline keywords and phrases that will help you focus on the passage. If you are unsure about a question, take an educated guess. Then move on. If you have time, review your answers. But don't second-guess yourself. These test-taking techniques will help you to maximize your performance on paragraph comprehension assessments.

Paragraph comprehension is an essential skill to succeed in the military. It allows military personnel to understand and apply written material, making informed decision and performing their duties efficiently. Individuals can improve their paragraph comprehension skills and do well on assessments such as the ASVAB by becoming familiar with the structure of the questions and the types, developing strategies to analyze and infer meaning, increasing reading speed and using effective test-taking methods. Mastering paragraph comprehension is essential to military success and overall readiness.

SECTION 3
MATH SKILLS

Chapter 5

Mathematics Knowledge And Operations

The Armed Services Vocational Aptitude Battery exam (ASVAB), which measures aptitude for military service, includes a section on mathematics. A good understanding of mathematics concepts, terminology and operations will help you achieve a high ASVAB math score. This chapter will examine the importance of mathematics operations and terminology in relation to ASVAB and offer strategies for improving your mathematical ability.

Math Terminology

The language of mathematics is a unique one, with its own vocabulary and terms. It is important to understand and use mathematical terminology correctly in order to solve mathematical problems on ASVAB. Test takers can navigate the different mathematical problems in the test by being familiar with terms like addition, subtraction and multiplication. The mastery of mathematical terms allows for clear communication of mathematical ideas and accurate representation.

Operation

Mathematical operations are the foundation of problem-solving and mathematical computations. The ASVAB tests the ability of candidates to perform mathematical operations, such as additions, subtractions and multiplications. You need to understand these operations in order to solve numerical problems efficiently and accurately.

Addition:

The operation of adding two or more numbers is to calculate their sum. Addition is used to solve problems involving adding quantities, finding sums or calculating amounts. Addition is used in the ASVAB to solve arithmetic questions, perform financial calculations or calculate the sum of a group of values. Understanding addition will help candidates calculate values accurately and give precise answers.

Subtraction:

Subtraction is the process of finding the difference between numbers. Subtraction is used in solving problems that require taking away quantities, determining changes, or measuring differences between values. Subtraction is a fundamental skill for solving word puzzles, analyzing numerical relationships and calculating differences. The ASVAB test measures candidates' abilities to perform subtraction efficiently and accurately.

Multiplication:

The process of multiplication is repeated addition. It is used to calculate the sum of two numbers or the product of groups of equal size. Multiplication skills are essential for solving problems involving rates, ratios, areas, or volumes. ASVAB questions often require multiplication skills in order to solve equations, analyze data or make predictions.

Division:

The division process is the act of dividing a quantity in equal parts, or of calculating the number of times a number is present in number. This method is used to solve problems that involve sharing, allocating resource, calculating rates or finding averages. Candidates who are proficient in division can solve proportionality and ratio problems, as well as interpret fractions. The ASVAB test evaluates the candidates' ability in performing division accurately and efficiently.

Mastering Operation for the ASVAB Exam

Candidates who want to excel on the ASVAB mathematics exam should develop a solid foundation in mathematical operations. Here are a few strategies that will help you improve your mathematical skills:

Practice Basic Operation:

It is important to practice addition, subtraction and multiplication and division regularly in order to improve accuracy and fluency. Solve different mathematical problems that involve these operations. This includes both numerical and verbal problems. Mental math and written methods can be used to practice calculations. You will become more efficient and faster at solving problems if you practice regularly.

Understanding the Properties of Operations

Learn the properties of operations such as commutative, associative, and distributives. These properties can be used to simplify calculations, reorder operations, or manipulate equations. Understanding these properties will allow you to look at problems in different ways and identify shortcuts.

Review and memorize key formulas and equations:

You may be asked to use specific equations or formulas to solve questions on the ASVAB. You should familiarize yourself with formulas that are commonly used in geometry, algebra and trigonometry. Apply these formulas to different contexts and reinforce your understanding.

Improve Mental Math Skills

You can improve your calculation speed and efficiency by developing mental math skills. Mental math techniques such as rounding numbers, estimating, and breaking them into smaller parts are all good to practice. You will be able to do quick calculations, estimate answers, and identify wrong answer choices during the test.

Solve real-world problems:

Solving real-world problems requiring the use of mathematical operations can help you improve your understanding of operations. Look for opportunities to use mathematical concepts and operations every day. Calculating expenses, analyzing data or solving practical questions related to distances, percentages or measurements can be part of this. A problem-solving mentality will help you identify and use the most relevant operations.

Use Online Resources to Study:

Online resources, study guides and practice tests designed specifically for the ASVAB test are available. These resources offer practice questions with detailed explanations to help you familiarize yourself with the type of problems and operations that will be tested on the ASVAB exam. Use these resources to reinforce your mathematical understanding and identify areas of improvement.

Fractions

In mathematics, fractions are frequently assessed by exams such as the ASVAB. Understanding fractions, and how to use them in operations is a crucial skill for success. This section will cover various aspects of fractions as well as strategies for mastering them.

Common Denominators - How to add and subtract fractions

When adding or subtracting fractions, it is important to use the same numerator. The denominator represents the total number, while the numerator represents the number of parts being considered.

Finding the least common denominator will allow you to add or subtract fractions with different denominators. You can then add or subtract numerators while maintaining the common denominator.

Adding Fractions:

Example: Add 3/5 and 1/4.

Step-by-step explanation:

Step 1: Find a common denominator.

The common denominator for 5 and 4 is 20.

Step 2: Rewrite the fractions with the common denominator.

3/5 = (3/5) * (4/4) = 12/20

1/4 = (1/4) * (5/5) = 5/20

Step 3: Add the numerators and keep the common denominator.

12/20 + 5/20 = (12 + 5)/20 = 17/20

Answer: 3/5 + 1/4 = 17/20

Subtracting Fractions:

Example: Subtract 2/3 from 7/8.

Step-by-step explanation:

Step 1: Find a common denominator.

The common denominator for 3 and 8 is 24.

Step 2: Rewrite the fractions with the common denominator.

7/8 = (7/8) * (3/3) = 21/24

2/3 = (2/3) * (8/8) = 16/24

Step 3: Subtract the numerators and keep the common denominator.

21/24 - 16/24 = (21 - 16)/24 = 5/24

Answer: 7/8 - 2/3 = 5/24

Remember, it's important to find a common denominator before adding or subtracting fractions. This allows for easier computation and ensures accurate results.

Multiplication and Reducing Fractions

Multiplying fractions is as simple as multiplying the numerators and denominators. The fraction that results is reduced to its simplest version by dividing the denominator and numerator by their largest common divisor. By reducing fractions, they become easier to understand and work with.

Multiplying Fractions:

Example: Multiply 2/5 by 3/4.

Step-by-step explanation:

Step 1: Multiply the numerators.

$2 * 3 = 6$

Step 2: Multiply the denominators.

$5 * 4 = 20$

Step 3: Simplify, if possible.

In this case, the fraction is already in its simplest form.

Answer: $(2/5) * (3/4) = 6/20$

Reducing Fractions:

Example: Reduce 24/36 to its simplest form.

Step-by-step explanation:

Step 1: Find the greatest common divisor (GCD) of the numerator and denominator.

The GCD of 24 and 36 is 12.

Step 2: Divide both the numerator and denominator by the GCD.

$24 \div 12 = 2$

$36 \div 12 = 3$

Step 3: Simplify the fraction.

The fraction 24/36 can be simplified to 2/3.

Answer: $24/36 = 2/3$

Remember, reducing a fraction involves dividing both the numerator and denominator by their greatest common divisor to obtain the simplest form.

Divide Fractions

Divide fractions by multiplying first fraction with reciprocal of second fraction. The reciprocal fraction is found by swapping the numerator with the denominator. Then multiply the numerator and denominator together to find the reciprocal. It is important to simplify any fractions that result from multiplication.

Example: Divide 2/3 by 4/5.

Step-by-step explanation:

Step 1: Take the reciprocal of the second fraction (the divisor).

The reciprocal of 4/5 is 5/4.

Step 2: Multiply the first fraction (the dividend) by the reciprocal of the second fraction.

(2/3) ÷ (4/5) is the same as (2/3) * (5/4).

Step 3: Multiply the numerators and denominators.

2 * 5 = 10 (numerator)

3 * 4 = 12 (denominator)

Step 4: Simplify, if possible.

In this case, the fraction 10/12 can be further simplified by dividing both the numerator and denominator by their greatest common divisor, which is 2.

10 ÷ 2 = 5

12 ÷ 2 = 6

Step 5: Simplify the fraction.

The fraction 10/12 can be simplified to 5/6.

Answer: (2/3) ÷ (4/5) = 5/6

Remember, when dividing fractions, you take the reciprocal of the second fraction and then multiply the fractions. Finally, simplify the resulting fraction if possible.

Converting Fractions into Mixed Numbers or Vice Versa

A fraction that is improper is one where the denominator is greater or equal to the numerator. A mixed number is a combination of a whole and a proper number. Divide the numerator and denominator to convert an improper fraction into a mixed number. The numerator is the fractional part of the whole number. Multiply the whole number by denominator, then add numerator. The new numerator is the result, while the denominator remains the same.

Converting Mixed Numbers to Improper Fractions:

Example: Convert 3 1/2 to an improper fraction.

Step-by-step explanation:

Step 1: Multiply the whole number by the denominator of the fraction.

3 * 2 = 6

Step 2: Add the product from Step 1 to the numerator of the fraction.

$6 + 1 = 7$

Step 3: Write the sum from Step 2 as the numerator and keep the denominator unchanged.

The denominator remains the same, which is 2.

Answer: 3 1/2 as an improper fraction is 7/2.

Converting Improper Fractions to Mixed Numbers:

Example: Convert 5/3 to a mixed number.

Step-by-step explanation:

Step 1: Divide the numerator by the denominator.

$5 \div 3 = 1$ with a remainder of 2.

Step 2: Write the whole number quotient as the whole number part of the mixed number.

The whole number part is 1.

Step 3: Write the remainder as the numerator of the fractional part.

The remainder is 2.

Step 4: Write the original denominator as the denominator of the fractional part.

The original denominator is 3.

Answer: 5/3 as a mixed number is 1 2/3.

Remember, when converting mixed numbers to improper fractions, multiply the whole number by the denominator and add the numerator. When converting improper fractions to mixed numbers, divide the numerator by the denominator and express the quotient as the whole number part, with the remainder as the numerator and the original denominator as the denominator of the fractional part.

How to express a fraction in other forms: Decimals and percents

You can also express fractions as decimals or percentages. Divide the numerator and denominator to convert a fraction into a decimal. The fraction will now be represented in decimal. Multiply the decimal equivalent by 100 to convert a fraction into a percentage. You will get the percentage equivalent of the fraction.

Converting Fractions to Decimals:

Example: Convert 3/4 to a decimal.

Step 1: Divide the numerator by the denominator.

$3 \div 4 = 0.75$

Answer: 3/4 as a decimal is 0.75.

Converting Fractions to Percents:

Example: Convert 5/8 to a percent.

Step 1: Divide the numerator by the denominator.

$5 \div 8 = 0.625$

Step 2: Multiply the result by 100 to get the percentage.

$0.625 * 100 = 62.5\%$

Answer: 5/8 as a percent is 62.5%.

Comparing Ratios

When comparing two or more quantities, ratios are used. You can express them as words, fractions or colons (e.g. "3 to 5"). The ratios help to solve problems that involve proportions or comparisons. If you have a 2:5 ratio, for example, it means that there are 2 units of a quantity and 5 units of a different quantity.

Example: Compare the ratios 3:4 and 5:6.

Step-by-step explanation:

Step 1: Write the ratios in fraction form.

3:4 is equivalent to 3/4.

5:6 is equivalent to 5/6.

Step 2: Find a common denominator for the fractions.

The common denominator for 4 and 6 is 12.

Step 3: Rewrite the fractions with the common denominator.

3/4 is equivalent to 9/12 (multiply numerator and denominator by 3).

5/6 is equivalent to 10/12 (multiply numerator and denominator by 2).

Step 4: Compare the fractions.

Since 9/12 is less than 10/12, we can conclude that the ratio 3:4 is less than the ratio 5:6.

Answer: The ratio 3:4 is less than the ratio 5:6.

It is important to solve a wide range of fraction problems in order to excel at fractions and the operations they involve on the ASVAB test. Addition, subtraction, multiplication, and division of fractions with various denominators are all important. Convert improper fractions to mixed numbers and decimals. Also, practice expressing fractions as percents and decimals. Also, become familiar with ratio concepts. Solve problems that require comparisons by using ratios.

You will be better prepared to answer questions about fractions on the ASVAB test if you have a good understanding of fractions. Regular practice and a solid grasp of fraction concepts will improve your performance and boost your confidence.

Writing in Scientific Notation

Scientific notation and roots are important mathematical concepts that are frequently assessed in various exams, including the ASVAB. Understanding and effectively utilizing scientific notation and roots can greatly enhance your mathematical fluency and problem-solving skills. In this chapter, we will delve into the significance of writing in scientific notation and exploring different types of roots, including perfect squares, irrational numbers, and other roots. We will also provide strategies to help you master these concepts for success on the ASVAB exam.

The scientific notation allows you to express very large numbers or very small ones in a concise and manageable way. It consists of a coefficient (a decimal number between 1 and 10) multiplied by a power of 10. Writing numbers in scientific notation allows for easier computation, comparison, and representation of quantities with a wide range of magnitudes.

To write a number in scientific notation, follow these steps:

- Identify the coefficient: Locate the first non-zero digit in the number and all the digits that follow.
- Determine the exponent: Count the number of places you moved the decimal point to obtain the coefficient. If the original number was greater than 1, the exponent is positive. If the original number was between 0 and 1, the exponent is negative.
- Write the number in the scientific notation format: Place the coefficient (with decimal point) followed by the letter 'x' and the exponent.

It is especially useful for calculations that involve very large or small numbers. This allows for a more efficient way to represent and manipulate these values.

Getting to the Root of the Problem:

Roots are mathematical operations that help us find the value that, when multiplied by itself a certain number of times, yields a given number. In this section, we will explore three types of roots: perfect squares, irrational numbers, and other roots.

Perfect Squares:

Perfect squares are numbers that are the result of multiplying an integer by itself. For example, 4, 9, and 16 are perfect squares because they can be expressed as 2^2, 3^2, and 4^2, respectively. Understanding perfect squares is essential for simplifying radicals, solving quadratic equations, and working with geometric shapes. Mastery of perfect squares enables quick mental calculations and facilitates problem-solving in various mathematical contexts.

Irrational Numbers:

Numbers that are irrational cannot be expressed in terms of a fraction, or as a ratio between two integers. These are non-repeating, non-terminating decimals. Examples of irrational numbers include $\sqrt{2}$, π (pi), and e (Euler's number). Irrational numbers play a significant role in geometry, trigonometry, and calculus, as they often arise in measurements, angles, and natural phenomena. Familiarity with irrational numbers allows for a deeper understanding of mathematical concepts and their real-world applications.

Other Roots:

Apart from perfect squares, there are other roots that involve finding the value that, when raised to a certain power, equals a given number. For instance, the cube root ($\sqrt[3]{}$) and the fourth root ($\sqrt[4]{}$) are types of other roots. These roots are used in various mathematical contexts, such as solving equations, working with geometric shapes, and simplifying expressions. Understanding and utilizing other roots expand your mathematical toolkit and enhance your problem-solving abilities.

Strategies to Master Scientific Notation and Roots:

To excel in writing numbers in scientific notation and exploring roots on the ASVAB exam, consider the following strategies:

– Practice Conversion to and from Scientific Notation: Work on converting numbers from standard notation to scientific notation and vice versa. Familiarize yourself with the rules and techniques for shifting the decimal point and determining the exponent. Regular practice will strengthen your proficiency and speed in writing numbers in scientific notation.

– Develop Mental Calculation Skills: Mental calculation is a valuable skill that allows for quick estimation and approximation. Practice mental calculations involving scientific notation, perfect squares, and other roots. This will help you develop a sense of numerical magnitude and improve your ability to perform calculations mentally.

– Study and Memorize Perfect Squares: Memorize the squares of integers from 1 to 15. This will enable you to recognize perfect squares quickly and simplify calculations involving square roots and other operations. Memorizing perfect squares saves time and reduces the likelihood of errors.

- Understand Properties of Irrational Numbers: Familiarize yourself with the properties and characteristics of irrational numbers. Study their relationship with fractions, decimals, and real numbers. Recognize when irrational numbers are likely to appear in problem-solving situations and practice working with them effectively.

- Apply Roots in Problem-Solving: Solve practice problems that involve roots, such as simplifying radicals, solving equations with radicals, and applying root operations in various mathematical contexts. Understand the connections between roots and their applications to develop a deeper understanding of their utility.

Mastering scientific notation and exploring different types of roots, including perfect squares, irrational numbers, and other roots, is crucial for success on the ASVAB exam and in various real-world applications. Writing numbers in scientific notation allows for efficient representation and computation of large and small quantities. Understanding perfect squares, irrational numbers, and other roots expands your mathematical toolkit and enhances your problem-solving abilities.

By practicing the conversion to and from scientific notation, developing mental calculation skills, studying perfect squares, understanding properties of irrational numbers, and applying roots in problem-solving, you can strengthen your mathematical fluency and boost your performance on the ASVAB exam. Furthermore, these skills will prepare you for military roles that require mathematical aptitude, critical thinking, and problem-solving skills. Embrace the power of scientific notation and roots to enhance your mathematical prowess and unlock new possibilities in your academic and professional journey.

Chapter 6

Algebra

A lgebra is a key tool in problem solving and critical thinking. It helps us represent relationships and solve equations. This chapter will introduce the basics of algebraic terminology and concentrate on solving equations. It is important to understand algebraic terms and solve equations in order to succeed on tests like the ASVAB. We will explore how to solve one-step addition, subtraction multiplication and division equations. We will also cover multi-step equations and simplifying equations. To make it easier to understand, we will provide valid examples.

Algebra Basic Terminology

Let's first learn some basic algebraic terms.

- Variables are used in algebra to represent unknown quantities or values. The letters x,y, and z are used to represent variables. They allow us express relationships between quantities, as well as solve equations.
- Constants can be defined as values that are constants in an expression or equation. Examples include 2, 5, or 3. In algebraic expressions, constants are used with variables.
- Coefficients multiply variables. In the expression 3x for example, the coefficient is 3. Coefficients are used to determine the magnitude or scale of variables in equations.

Solving Equations

Equations are statements which assert the equality of expressions. In order to solve equations, you must determine the values of variables which satisfy the equation. We will explore the different types of equations, and how to solve them.

Solving one-step equations involving addition and subtraction:

In one-step equations, there is only one operation: either addition or subtraction. It is important to isolate the variable from one side of an equation using the inverse operation. You can use the following example:

Solve the equation: $2x + 5 = 11$

- – Subtract 5 from each side to isolate the variable: $2x = 6$
- – Divide the two sides by 2, to find x. $x = 3$.

Solving One Step Equations Involving Division and Multiplication:

Multiplication and division one-step problems are similar to addition and subtraction one-step problems. However, they require inverse operations in order to isolate the variables. As an example:

Solve the equation: $4x/2 = 6$

- – Multiply both sides of the variable by 2. $4x = 12$
- – Divide the two sides by four to find x. $x = 3$.

Solving Multistep Equations

Multistep equations require multiple operations to isolate the variable. To simplify the equation, follow the order of operations. You can use the following example:

Example 3: Solve $2x + 3 - 5 = 10$.

- Combining like terms, simplify the equation as follows: $2x + 2 = 10$.
- Add 2 on both sides of the variable to isolate it: $2x = 12$
- Divide the two sides by 2, to find x. $x = 6$.

Simplifying Equations

To simplify equations, you combine terms that are similar and apply properties of operations in order to make it easier to manage. As an example:

Example 4: Simplify $3(x + 2) = 2(x - 4) = 7$.

- – Use the distributive property to solve for $3x + 6 + 2x + 8 = 7$
- – Add like terms together: $x + 7 =$
- – Add 14 to both sides of the variable $x = -7$.

Use FOIL

It is used to multiply binomials. It is an acronym for First, Inner, Outer and Last. This represents the order of multiplying the terms. As an example:

Example 5: Multiplying $(x + 2)$ using the FOIL method.

- First, multiply the first terms. $x * x = x2$.
- Outer: Multiply outer terms as follows: $x * 3 = -3x$.
- Inner: Multiply inner terms as follows: $2 * 2 = 2x$.
- Last: Multiplying the last terms: $2 * -3 = 6$

Combining the terms: $x2 + x - 6$.

It is important to master algebraic terms and solve equations in order to succeed on tests like the ASVAB. This will also help you develop problem-solving abilities. Understanding variables, constants and coefficients is a good foundation for algebraic equations and expressions. You can navigate algebraic problems with confidence by using strategies to solve multistep equations and one-step equations. The FOIL method will also improve your ability to multiplication binomials.

Tackling Two-Variable Equations:

Substitution: In this method, you solve one equation for one variable and substitute that expression into the other equation. For example, consider the equations:

$x + y = 8$ and $2x - y = 2$.

By solving the first equation for x $(x = 8 - y)$ and substituting it into the second equation, you get:

$2(8 - y) - y = 2$.

Simplifying further, you can solve for y, and then substitute the value of y back into one of the original equations to find x.

Combining equations: In this method, you manipulate the equations in a way that eliminates one variable when the equations are added or subtracted. For example, consider the equations:

$3x + 2y = 10$ and $2x - 4y = 16$.

By multiplying the second equation by 2 and then adding the two equations, you can eliminate y:

$(3x + 2y) + (4x - 8y) = 10 + (2 * 16)$.

Simplifying further, you can solve for x, and then substitute the value of x back into one of the original equations to find y.

Explaining Exponents in Algebra:

Exponents are used to represent repeated multiplication of a base number. They are written as a superscript number after the base. For example:

x^2 means $x * x$.

3^4 means $3 * 3 * 3 * 3$.

$(a + b)^3$ means $(a + b) * (a + b) * (a + b)$.

Exponents follow specific rules, such as:

Multiplying two numbers with the same base: $x^a * x^b = x^{(a+b)}$.

Dividing two numbers with the same base: $x^a / x^b = x^{(a-b)}$.

Raising a power to another power: $(x^a)^b = x^{(a*b)}$.

These rules help simplify algebraic expressions and solve equations more easily.

Factoring Algebra Expressions to Find Original Numbers:

Pulling out the greatest common factor: When factoring an expression, you look for the largest factor that divides evenly into all the terms. For example:

In the expression $4x + 8$, the greatest common factor is 4, so you can factor it as $4(x + 2)$.

Factoring a three-term equation: When you have a quadratic expression with three terms, you look for two binomials that multiply to give you the original expression. For example:

In the expression $x^2 + 5x + 6$, you can factor it as $(x + 2)(x + 3)$, which gives you two binomials that, when multiplied, result in the original expression.

Solving Quadratic Equations:

Quadratic equations are equations in the form $ax^2 + bx + c = 0$, where a, b, and c are constants. To solve these equations, you can use methods like:

Factoring: If the quadratic expression can be factored, you can set each factor equal to zero and solve for the variable.

Quadratic formula: The quadratic formula is used when factoring is not feasible. It states that for any quadratic equation $a^2 + bx + c = 0$, the solutions for x are given by $x = (-b \pm \sqrt{(b^2 - 4ac)})/(2a)$.

Solving Inequalities:

Inequalities compare two expressions and show the relationship between them using symbols such as < (less than), > (greater than), ≤ (less than or equal to), ≥ (greater than or equal to). To solve inequalities, you follow similar rules as in solving equations, but with some modifications:

When multiplying or dividing both sides of an inequality by a negative number, you must reverse the inequality symbol.

When adding or subtracting a number from both sides of an inequality, the inequality remains the same.

For example, to solve the inequality $2x + 3 > 7$, you subtract 3 from both sides to get $2x > 4$. Then, you divide both sides by 2 to find $x > 2$ as the solution.

Apply algebraic concepts in real-world situations and practice solving different types equations. These examples will help you to understand and apply the concepts. Use algebra to improve your math fluency, problem-solving skills, and critical thinking in academic and professional settings.

Chapter *7*

Geometry

Geometry is a branch of mathematics that deals with the properties, relationships, and measurements of shapes, sizes, and positions of figures. It encompasses various concepts and formulas that help us understand and solve problems related to the geometric world. In this explanation, we will delve into the following topics in geometry: Perimeter and Area, Angles, Triangle Types, Quadrilaterals, Circles, Calculating Volume, Surface Area of Solids, Breaking Down Combined Figures, Mapping Out Coordinates, and Helpful Geometry Formulas.

Perimeter and Area:

Perimeter refers to the distance around the outside of a shape, while area represents the space inside the shape. The formulas for calculating the perimeter and area of common geometric figures are as follows:

Perimeter:

Rectangle: Perimeter = 2 * (length + width)

Square: Perimeter = 4 * side length

Triangle: Perimeter = sum of all three side lengths

Circle: Perimeter = 2 * π * radius

Area:

Rectangle: Area = length * width

Square: Area = side length^2

Triangle: Area = (base * height) / 2

Circle: Area = π * radius^2

Angles:

Angles are formed when two rays share a common endpoint called a vertex. The size of an angle is measured in degrees. Here are some important concepts related to angles:

Parallel Lines: Parallel lines are lines that never intersect and are always the same distance apart. When a line intersects two parallel lines, it creates pairs of corresponding angles, alternate interior angles, and alternate exterior angles.

Naming Angles: Angles can be named based on their location and size. Some common angle names include:

Acute angle: An angle that measures less than 90 degrees.

Right angle: An angle that measures exactly 90 degrees.

Obtuse angle: An angle that measures greater than 90 degrees but less than 180 degrees.

Straight angle: An angle that measures exactly 180 degrees.

Complementary angles: Two angles that add up to 90 degrees.

Supplementary angles: Two angles that add up to 180 degrees.

Triangle Types:

Triangles are three-sided polygons. They can be classified into different types based on the length of their sides and the measure of their angles:

Scalene Triangle: A scalene triangle has no equal sides and no equal angles.

Isosceles Triangle: An isosceles triangle has two equal sides and two equal angles.

Equilateral Triangle: An equilateral triangle has three equal sides and three equal angles measuring 60 degrees each.

Right Triangle: A right triangle has one angle that measures 90 degrees.

Obtuse Triangle: An obtuse triangle has one angle that measures greater than 90 degrees.

Acute Triangle: An acute triangle has all three angles measuring less than 90 degrees.

Quadrilaterals:

Quadrilaterals are four-sided polygons. They come in various forms, and some common types include:

Square: A square is a quadrilateral with four equal sides and four right angles.

Rectangle: A rectangle is a quadrilateral with opposite sides that are equal and four right angles.

Parallelogram: A parallelogram is a quadrilateral with opposite sides that are parallel.

Rhombus: A rhombus is a quadrilateral with four equal sides.

Trapezoid: A trapezoid is a quadrilateral with one pair of parallel sides.

Circles:

Circles are perfectly round shapes defined by a curved line called the circumference. Here are some key concepts related to circles:

Circumference: The circumference of a circle is the distance around its outer boundary. It can be calculated using the formula: Circumference = 2 * π * radius.

Area of a Circle: The area of a circle is the measure of the space enclosed by its boundary. It can be calculated using the formula: Area = π * radius^2.

Calculating Volume:

Volume refers to the amount of space occupied by a three-dimensional object. The formulas for calculating the volume of common shapes are as follows:

Cube: Volume = side length^3

Rectangular Prism: Volume = length * width * height

Cylinder: Volume = π * radius^2 * height

Cone: Volume = (1/3) * π * radius^2 * height

Sphere: Volume = (4/3) * π * radius^3

Surface Area of Solids:

Surface area is the total area of all the faces of a three-dimensional object. The formulas for calculating the surface area of common shapes are as follows:

Cube: Surface Area = 6 * side length^2

Rectangular Prism: Surface Area = 2 * (length * width + length * height + width * height)

Cylinder: Surface Area = 2 * π * radius^2 + 2 * π * radius * height

Cone: Surface Area = π * radius * (radius + slant height)

Sphere: Surface Area = 4 * π * radius^2

Breaking Down Combined Figures:

Sometimes, geometric figures can be composed of multiple simpler shapes. To find the area or perimeter of such combined figures, you can break them down into smaller shapes and then sum up their individual areas or perimeters.

Mapping Out Coordinates:

Coordinates are used to specify the position of points on a plane. The coordinate system consists of an x-axis and a y-axis, intersecting at a point called the origin (0, 0). Each point is represented by an ordered pair (x, y), where x denotes the horizontal distance from the y-axis, and y denotes the vertical distance from the x-axis.

Helpful Geometry Formulas:

Here are some additional formulas that can be useful in solving geometry problems:

Pythagorean Theorem: In a right triangle, the square of the length of the hypotenuse (the side opposite the right angle) is equal to the sum of the squares of the other two sides. It can be written as $a^2 + b^2 = c^2$.

Law of Sines: In any triangle, the ratio of the length of a side to the sine of its opposite angle is constant. It can be written as $\sin(A)/a = \sin(B)/b = \sin(C)/c$.

Law of Cosines: In any triangle, the square of the length of one side is equal to the sum of the squares of the other two sides minus twice their product, multiplied by the cosine of the included angle. It can be written as $c^2 = a^2 + b^2 - 2ab * \cos(C)$.

Area of a Triangle using Heron's Formula: For a triangle with sides of lengths a, b, and c, the area can be calculated using the formula: Area = $\sqrt{s * (s - a) * (s - b) * (s - c)}$, where s represents the semi-perimeter (s = (a + b + c)/2).

In summary, geometry encompasses various concepts and formulas that allow us to understand the properties and relationships of shapes and figures. From calculating perimeters and areas to analyzing angles, triangles, quadrilaterals, circles, and three-dimensional objects, these concepts provide a foundation for solving geometry problems and understanding the world of shapes and measurements.

Chapter 8

Arithmetic Reasoning: Math Word Problems

A rithmetic reasoning is the use of mathematical operations to solve word problems in real life. You can use a step-by-step approach to solve these problems.

Start by reading the entire problem. Be sure to understand the context of the problem and identify key information.

Determine the problem or question that is being asked. Does the question ask for a particular value, a comparison or a calculation. Understanding the question helps you to focus on the information that is relevant and determine the best approach.

Find the facts. Identify all the pertinent facts and numbers that are given in the question. Write down or highlight any information important to solving the problem. Attention should be paid to the units of measurement, timelines, and other details which may impact the solution.

Transform the problem or set of problems into a mathematical formula. Set up the problem using the information provided and identify all variables. Apply the appropriate mathematical operations to the equation (such as addition or subtraction, division, multiplication or multiplication) and find the solution.

In certain cases, drawing diagrams can help you visualize the problem. They also provide more clarity. Diagrams are particularly helpful for problems involving geometry or spatial reasoning. Use the diagrams to visualize and represent the information.

Following these steps will improve your ability to solve arithmetic problems. You will improve your problem-solving skills by practicing with different types of word problems.

Mastering mathematics knowledge and operations is vital for achieving success on the ASVAB exam. A solid understanding of mathematical terminology and proficiency in performing operations like addition, subtraction, multiplication, and division are key to solving mathematical problems accurately and efficiently. By practicing basic operations, understanding their properties, memorizing key formulas and equations, enhancing mental math skills, solving real-world problems, and utilizing online resources, candidates can strengthen their mathematical proficiency and improve their performance on the ASVAB exam. Developing strong mathematical skills not only increases your chances of obtaining a high score on the exam but also prepares you for various military roles that require mathematical aptitude and problem-solving abilities.

SECTION 4
GENERAL SCIENCE

Chapter 9

General Science And Life Science

Understanding Forms Of Measurement

The measurement of things is an important part of understanding the world and our own lives. It provides us with useful information and insight by allowing us to assess and quantify different quantities or attributes. The different types of measurement include a variety of methods and units that serve a purpose.

Length is a common way to measure. The length is the distance or extent between two points. Understanding length helps us to understand spatial relationships, whether we're measuring the height of the building, the width or length of the road, or a piece fabric. To express length, units such as meters or feet are often used. This allows for accurate calculations and comparisons.

Weight is another important form of measurement. The weight is used to determine the weight of an object. Weight is important in many contexts, from weighing ingredients for a recipe to assessing a vehicle's mass. Weight is measured in kilograms, pounds or ounces. This allows us to compare the relative weight of objects.

The measurement of time is also important. It helps us understand the sequence or duration of events. Time measurement is essential for scheduling appointments, tracking your progress or understanding historical timelines. We can express time using units like seconds, hours, minutes and years.

It is essential to understand and quantify thermal conditions. It allows us to determine the level of warmth or coldness in objects, substances or environments. Temperature measurement is important for many reasons, including setting the thermostat at home, monitoring weather patterns or ensuring that machinery works properly. Temperature is commonly expressed in units such as Fahrenheit or Kelvin, which provide a consistent scale.

There are many specialized measurement forms, which cater to certain fields or disciplines. In the field of economics, for example, financial measurements like income, expenses and profit are essential to assessing economic health. In the sciences, measurements are crucial to disciplines such as physics, biology, and chemistry. They help us understand physical properties, chemical processes, and biological processes.

Forms of measurement are essential tools for understanding. They enable us to compare, analyze, and quantify attributes, quantities, or phenomena in a meaningful and structured way. We gain insight into our environment by using different units and measurement methods. This variety of fields.

Biology

Biology is a scientific field that studies living organisms and their interactions with each other and the environment. It encompasses a wide range of sub-disciplines, including ecology and biodiversity. Ecology is the study of the relationship between organisms and the environment in which they live, while biodiversity is the diversity and abundance of all living organisms. Additionally, categorizing Mother Nature involves classifying and organizing the diverse forms of life on Earth. Here, we will explore these three interconnected topics, delving into their significance and exploring the methods and concepts used in each.

Ecology:

Ecology is the study of the relationships between organisms and their environment. It aims to understand the distribution and abundance of organisms, the flow of energy and nutrients through ecosystems, and the factors that shape ecological communities. Ecological research provides crucial insights into the functioning of ecosystems and the impacts of human activities on natural systems.

One key concept in ecology is the ecosystem. An ecosystem consists of a community of organisms interacting with each other and with their physical environment. It includes both biotic (living) and abiotic (non-living) components, such as plants, animals, soil, water, and climate. Ecologists study the structure and dynamics of ecosystems, investigating how energy and matter flow through different trophic levels and how organisms adapt to their environment.

Ecological research also examines various ecological processes, such as nutrient cycling, predation, competition, and symbiosis. These processes shape the composition and functioning of ecosystems, influencing the distribution and abundance of species. Furthermore, ecologists' study

ecological succession, which is the gradual change in species composition and community structure over time. Succession can occur in both terrestrial and aquatic ecosystems and is influenced by factors such as disturbance, climate, and species interactions.

Another important area of study within ecology is population ecology. Population ecology focuses on the dynamics of populations, including factors that affect population size, growth, and density. It explores concepts such as birth rates, death rates, immigration, emigration, and the carrying capacity of an environment. Population ecologists employ mathematical models to understand and predict population changes, aiding in conservation efforts and wildlife management.

Biodiversity:

The term biodiversity refers to a wide range of organisms that can be found on Earth or in particular areas. The term encompasses species diversity, genetic diversity in species and ecosystem diversity. Biodiversity plays a vital role in the functioning and stability of ecosystems, and it provides many ecosystem services to humans.

Species diversity, a component of biodiversity, refers to the amount of different species available in an area and their relative abundance. High species diversity is indicative of a healthy and resilient ecosystem. It ensures the presence of a variety of ecological niches and promotes stability in the face of environmental changes.

Genetic diversity, on the other hand, focuses on the genetic variation within a species. Genetic diversity is crucial for the adaptation and survival of populations. It provides the raw material for evolutionary processes, allowing species to respond to environmental pressures and reducing the risk of extinction. Conservation efforts often emphasize the preservation of genetic diversity, as it contributes to the long-term viability of species.

Ecosystem diversity refers to the variety of ecosystems present in a region or on a larger scale. Ecosystems differ in terms of their structure, composition, and functioning. Each ecosystem has unique characteristics, such as the types of plants, animals, and physical features present. Ecosystem diversity ensures a range of ecological processes and services, such as carbon sequestration, water purification, and nutrient cycling.

Loss of biodiversity is a serious threat to human and ecosystem health. The rapid decline of biodiversity has been caused by human activities such as habitat destruction and pollution. This loss of biodiversity can disrupt ecosystem functions, reduce ecosystem resilience and reduce the availability ecosystem services. Conservation efforts are aimed at reducing biodiversity loss and promoting sustainable management practices in order to protect the Earth's biological diversity.

Categorizing Mother Nature

Categorizing Mother Nature involves the classification and organization of the diverse forms of life on Earth. Taxonomy is the branch of biology that focuses on identifying, naming, and classifying organisms based on their characteristics and evolutionary relationships. The classification system developed by Carl Linnaeus in the 18th century forms the foundation of modern taxonomy.

Taxonomy classifies organisms into a hierarchical system of categories, known as taxa. The major taxa, from broadest to most specific, are domain, kingdom, phylum, class, order, family, genus, and species. This hierarchical structure reflects the evolutionary relationships between organisms. The two-part scientific name assigned to each species, consisting of the genus and species epithet, is known as binomial nomenclature.

Advances in molecular biology and genetic techniques have revolutionized the field of taxonomy. DNA sequencing allows scientists to examine the genetic material of organisms and compare their genetic relatedness. This molecular data complements traditional morphological and anatomical characteristic used in taxonomy, providing a more comprehensive understanding of evolutionary relationships.

Phylogenetics is a field within taxonomy that reconstructs the evolutionary history of organisms, known as phylogeny. Phylogenetic trees depict the branching patterns of lineages and help elucidate the shared ancestry and relationships among species. Phylogenetics aids in resolving taxonomic uncertainties and assists in understanding the evolutionary processes that have shaped life on Earth.

Modern taxonomic approaches also consider the concept of evolutionary classification. Evolutionary classification groups organisms based on their evolutionary history and shared ancestry. This approach provides insights into the patterns and processes of evolution and highlights the interconnectedness of all living organisms.

In addition to traditional taxonomy, other systems of classification exist to categorize biodiversity. For example, ecological classification categorizes organisms based on their ecological roles and adaptations. Functional traits, such as feeding habits, reproductive strategies, and habitat preferences, are considered in ecological classification systems. These classifications focus on the ecological functions and interactions of organisms within ecosystems, providing valuable insights into ecosystem dynamics.

Biology encompasses various sub-disciplines that help us understand the intricate web of life on Earth. Ecology investigates the relationships between organisms and their environment, providing insights into the functioning of ecosystems and the impacts of human activities. Biodiversity focuses on the variety and abundance of living organisms, emphasizing the importance of preserving species, genetic, and ecosystem diversity. Categorizing Mother Nature involves classifying and organizing the diverse forms of life on Earth, allowing us to better understand the evolutionary relationships between organisms. Together, these fields of study contribute to our knowledge of the natural world and inform conservation efforts to safeguard the Earth's biological heritage.

The human body is an amazing and complex organism made up of many interconnected systems which work together to maintain life. Each system is unique and has a specific function. They all play a vital role in the health and overall functioning of the human body. This essay will examine the intricate workings of major body systems including the nervous, respiratory, digestive, musculoskeletal, endocrine, lymphatic, integumentary, reproductive, and musculoskeletal systems.

The Human Body Systems

The Nervous System

The nervous system controls and coordinates various bodily functions. The nervous system is composed of two major components: the central nerve system (CNS), including the brain and spinal chord, and the peripheral neural system (PNS), consisting of nerves and other ganglia which connect the CNS with the rest of body.

Nerve impulses are electrical signals that transmit information through the nervous system. The basic units of nervous system are neurons, which receive and transmit impulses. Sensory neurons collect information about the environment, relaying it to the CNS. Motor neurons then transmit the signals from the CNS directly to muscles and glands to enable movement and response.

The brain is the central organ of our nervous system. It controls and integrates all bodily functions including memory, thoughts, feelings, and sensory perception. The spinal cord is a communication path between the brain, the rest of the human body and facilitates reflexes. It also transmits sensory and motor information.

The Heart, Blood, and Circulatory Systems:

The circulatory system includes the blood vessels, heart and blood. It transports oxygen, hormones and nutrients throughout the body. The heart is a muscular motor that pumps blood through capillaries, veins and arteries.

The heart is made up of four chambers, two atria, and two ventricles. The heart undergoes rhythmic movements called cardiac cycles that result in blood pumping. The circulatory system consists of two parts: the systemic circulation that delivers oxygenated body tissue and the pulmonary system which transports the deoxygenated blood into the lungs to be oxygenated.

The blood, a special connective tissue, performs essential functions. White blood cells protect against pathogens, while red blood cells transport oxygen. Platelets help to clot blood and prevent excessive bleeding. Plasma, a liquid component of the blood, transports hormones and waste products.

Renal 9-1-1 Emergency Toxin Filtering

The kidneys, ureters and bladder are all part of the renal system. They play a crucial role in maintaining fluid equilibrium, regulating electrolytes and eliminating waste from the body. The kidneys filter the blood and produce urine as a waste.

Nephrons are the functional units that filter blood and produce urine. Nephrons maintain the pH, water, and electrolyte balance by excreting waste and reabsorbing essential substances. The urine produced in the kidneys is temporarily stored in the bladder by the ureters before it is eliminated through the urethra.

The Respiratory System

The respiratory system is responsible for the exchange of gases that occurs between the body, the air and the surrounding environment. The respiratory system involves oxygen intake and carbon dioxide elimination. The respiratory system is made up of the lungs and airways.

Air enters the lungs through the mouth or nose during inhalation. Oxygen diffuses through the tiny air sacs, or alveoli in the lungs. Exhalation is the process by which carbon dioxide, a by-product of cell metabolism, is removed.

The Digestive System

The digestive system is responsible for processing food, extracting nutrients and eliminating waste products. The digestive system is made up of several organs including the stomach, oesophagus and small intestine.

In the mouth, food is mechanically broken by chewing. Saliva then mixes with it. The process continues in the stomach where gastric acids further break down food. Villi, which are specialized structures in the small intestine, facilitate the absorption of nutrients. The large intestine absorbs electrolytes and water, resulting in faeces.

The liver is crucial in digestion because it produces bile that helps in fat breakdown and absorption. The gallbladder releases and stores bile while the pancreas creates digestive enzymes to break down carbohydrates and proteins.

You can't Muscling your Way Through Life

It provides the body with structure, movement, and support. This system includes bones, muscles and tendons.

The bones are the structure of the body, protecting vital organs. The joints connect them to one another, allowing movement. Tendons attach muscles to bones, which contract and produce force. This allows movement. Ligaments provide stability by connecting bones. These components work together to allow a variety of movements, ranging from simple ones like walking, to more complex ones like sports.

Take the Skeletons out of the Closet

The skeletal system is made up of bones, cartilage and other connective tissue. It performs several vital functions including supporting, protecting, facilitating movement and producing blood.

The four main types of bones are: flat, long, and irregular. The bones are made up of living cells that are embedded in a matrix mineralized, which provides strength and resilience. Cartilage is a flexible connective tissue that cushions joints and supports structural integrity. The skeletal system stores minerals such as calcium, phosphorus and bone marrow where blood cells are made.

The Endocrine system:

Endocrine system regulates body functions by secreting hormones. These chemical messengers travel through bloodstream to target cells. The endocrine system includes many glands such as the thyroid gland, pituitary, adrenal, pancreas and reproductive glands.

The pituitary, which is located at the base the brain, regulates other endocrine organs and produces hormones to control growth, reproduction and metabolism. The thyroid gland produces hormones which regulate metabolism and energy balance. The adrenal glands secrete hormones that are involved in the stress response as well as salt and water balance.

The pancreas is responsible for producing insulin and glucagon hormones, which regulate blood sugar. The reproductive glands such as the testes of males and ovaries of females produce sex-hormones that play a part in reproduction and sexual growth.

The Lymphatic System

The lymphatic system is made up of lymphatic nodes and organs such as thymus and spleen. It is responsible for fluid balance, immunity, and absorption of dietary fatty acids.

Lymphatic vessels transport lymph, a fluid that is derived from the interstitial liquid, throughout the entire body. The lymph nodes remove pathogens and other foreign substances from the lymph. The spleen filters the blood and removes pathogens and old red blood cell. The thymus is crucial in the maturation and development of T cells. These are important components of our immune system.

The Integumentary System

The integumentary barrier, which is composed of hair, skin, nails and glands, protects us from the outside world. It controls body temperature, protects against dehydration and is involved in sensory perception.

The skin is made up of three layers, the epidermis (outermost layer), dermis (dermis), and hypodermis. The epidermis provides protection and waterproofing. Dermis is made up of blood vessels, sweat glands, nerve endings and hair follicles. The subcutaneous layer of the hypodermis contains fat cells that help insulate your body.

The Reproductive System

The reproductive system is the one responsible for producing offspring. It includes the testes in males, as well as the epididymis and vas deferens. It includes the ovaries and fallopian tube, as well as the uterus, vagina, cervix and breasts in females.

The testes of males are the source of sperm, and also male hormones. The epididymis is where sperm matures. They are then transported by the vas deferens through ejaculation. Seminal fluid is produced by glands, including the prostate and seminal versicles.

The ovaries in females produce female hormones and eggs. During ovulation an egg is released by the ovary. It travels down the fallopian tube to the uterus. The fertilized egg will implant in the uterus if fertilization takes place. The uterus is the place where a foetus develops, and the cervix links the uterus with the vagina. Breasts are responsible for producing milk to nourish newborns.

Keep the body in peak operating condition:

For the body to function at its best, it requires proper nutrition, adequate hydration and rest, regular exercise and a healthy lifestyle. Nutrition is essential for growth, development and cellular functions. Hydration is essential for supporting body functions and maintaining fluid balance.

Rest and sleep is necessary for mental and physical rejuvenation. Regular exercise improves cardiovascular health and flexibility. Maintaining optimal mental and physical health requires a balanced lifestyle that includes stress management, social connections and managing stress.

Human body systems are interconnected to ensure survival and wellbeing. Each system is vital. From the nervous system's control and coordination to the heart and circulatory systems' pumping action, to the filtration and exchange of gases by the respiratory system and the kidney system, to the digestion and absorption nutrients by the digestive system and musculoskeletal, each system has a role to play.

Endocrine, lymphatic and reproductive systems are involved in hormone regulation, immunity defence, protection and reproduction functions. Maintaining peak operating condition also requires a holistic lifestyle that includes proper nutrition, hydration and rest.

Understanding the intricate systems of the body allows us to appreciate our bodies' remarkable resilience and complexity. The book emphasizes the importance to take care of your physical and mental health and the interconnectedness between the various body components.

Plant Physiology

Plant physiology, a branch of biology, is concerned with the study of the way plants function, including the processes that are essential to their growth, development and survival. Plants play an important role in the ecosystem, and they provide many benefits for humans and other organisms.

Here, we will explore the fascinating worlds of plant physiology. We will examine the types of plants and where they grow, the reproduction of plants, the structure of the cells and how the cell processes are used for plant growth.

Types of plants:

The plants are a diverse group of species, with a variety of characteristics and adaptations. Plants can be classified based on a variety of criteria. Plants are classified according to their life cycle, with annuals, biennials and perennials.

Annual plants finish their life cycle in one year. They germinate from seeds, grow, flower, produce seeds, and then die. Biennial plants live for two years, the first being vegetative growth and the second flowering and producing seeds. Perennial plants can live for many years and regenerate and grow each year. They can reproduce by seeds, bulbs or rhizomes.

Plants are also classified based on the habitats and adaptations they have. Water lilies are adapted for aquatic environments while cacti, succulents, and other plants thrive in arid climates. Other plants have adaptations that are specific to their habitat, like epiphytes which grow on other plants and do not require soil nutrients.

Where plant processes take place:

The plant processes are carried out in different parts, with each part having a specific function and role. The root system anchors the plant to the soil, and absorbs water and nutrients. The root hairs are extensions from the root surface. They increase the surface for absorption.

The stem, leaves and reproductive structures make up the shoot system. The stem is responsible for supporting the plant, transporting water, minerals and sugars and serving as a place to attach leaves and reproductive organs. The leaves are the primary sites for photosynthesis. Here, sunlight is transformed into chemical energy. Reproductive structures such as flowers produce seeds and allow plant reproduction.

Plant Reproduction:

Sexual and asexual reproduction are essential to the survival of a species. The fusion of both male and female gametes results in the formation seeds. Angiosperms have flowers as reproductive structures. Pollination occurs when pollen from male reproductive organs is transferred to female reproductive organs. Fertilization results in the formation of seeds that can be dispersed to germinate and form new plants.

Asexual reproduction is a form of plant reproduction that does not require the fusion of gametes. It produces genetically identical offspring. It can be done in a variety of ways, including vegetative propagation. This is when new plants are created from stems, leaves, roots or other specialized plant parts. Asexual reproduction is also possible in some plants through the production of special structures such as tubers, runners, or bulbs.

Take a look at the cells:

Plant cells are unique in that they can perform specific functions. Cells are fundamental units of the life cycle. Plant cells are eukaryotic cellular structures that have a membrane-bound nucleus.

Cell Structure

Plant cells contain several organelles with distinct functions. Cell walls are rigid structures that surround the membrane of the cell, offering support and protection. The cell wall is composed primarily of cellulose. This complex carbohydrate is the main component. The cell membrane or plasma membrane regulates substances entering and leaving the cell.

The nucleus is the site of photosynthesis, containing chlorophyll pigments that capture sunlight and convert it into chemical energy. The chloroplasts, which contain pigments called chlorophyll that convert sunlight into chemical energy, are where photosynthesis occurs. Mitochondria produce energy by cellular respiration.

Sorting and packaging proteins is done by the Golgi apparatus. The Golgi apparatus also modifies, sorts and packages proteins. Waste, water and nutrients are stored in vacuoles.

Profiting from cell processes:

Plant cells use a variety of processes to perform essential functions for growth and survival. Photosynthesis is an essential process by which plants convert light, carbon dioxide and water into oxygen and glucose. Chloroplasts containing chlorophyll capture sunlight energy which is then used to synthesize stored energy in the form of glucose.

Cellular respiration occurs when cells breakdown glucose to release energy (ATP, adenosine Triphosphate). Mitochondria are crucial in this process. They convert glucose and oxygen to carbon dioxide, water and ATP.

Plant cells are also involved in other processes, such as transpiration. This is the process of releasing water vapor through stomata from leaves, which helps regulate temperatures and transport nutrients. Plant cells also undergo cell division to allow for growth, repair and the formation new tissues.

Plant physiology explores the complex processes and functions that plants perform. Understanding plant types, the location of plant processes, the cell structure and how cells are used, as well as the reproduction of plants and their cell structure, can provide insights into plants' remarkable abilities to adapt, grow and survive in different environments. Plants are essential to the ecosystem. They provide oxygen, food and shelter for many organisms. Plants are important to human health, as they provide food, medicine and materials. Studying plant physiology allows us to gain a greater understanding of the resilience and complexity of plants. We also recognize the importance of protecting and preserving these vital organisms, for present and future generations.

Genetics

The branch of biology known as genetics focuses on studying genes, heredity and variation within living organisms. The essay explores how traits are passed down, the role that genes play in our personality, and the transmission genetic information. This essay will explore the fascinating world genetics. We will cover topics like gene coverage, sex-determination, and the transmission of genes within families.

Covering Genes:

The instructions that are contained in genes (deoxyribonucleic acids) for the building and maintenance of an organism can be found within these segments of DNA. Genes are the blueprints for producing proteins that are vital for the structure and functionality of the cells and body. Gene coverage is the idea of identifying and describing all genes in an organism's DNA.

The Human Genome Project was completed in 2003. It was a groundbreaking scientific project that sought to map and sequence the human genome. Scientists identified approximately 20,000 to 25,000 genes within the human genome through this project. As scientists continue to study the functions and interactions between all genes in different organisms, gene coverage is a very active research area.

How to determine your gender with two letters:

The process of determining an individual's gender is called sex determination. Sex is determined in most species including humans by the presence or absence of specific chromosomes. In humans, the females usually have two X-chromosomes.

When reproducing, the egg of the mother will always carry an X-chromosome. Sperms from the father, however, can either have an X- or a Y-chromosome. The resulting zygote (XX) will be a female if an X-carrying sperm fertilizes the egg. If a Y carrying sperm fertilizes an egg, the resulting zygote develops into a female (XX).

The sex determination is controlled by genes on chromosomes. The SRY gene (sex-determining area Y), located on the human Y chromosome triggers the development male characteristics in embryonic development. Understanding the molecular mechanism of sex determine has wide implications for various fields including reproductive medicine and evolution biology.

Know which genes are passed down the family line:

Inheritance of genes in families is governed by specific patterns. Mendelian Genetics is the name given to the fundamental principles of inheritance first described by Gregor Mendel during his lifetime in 19th century. Mendel's pea plant experiments demonstrated dominant and recessive characteristics, as well as the concept of segregation.

Three main inheritance patterns exist: autosomal dominance, autosomal regressive and sex linked. In autosomal dominance inheritance, one copy of a mutant gene from a parent is enough to manifest the disorder or trait. Huntington's and Marfan syndrome are examples.

For autosomal recessive inheritance to occur, both copies of a gene (one from each parent) must be mutated. Cystic fibrosis, for example, and sickle-cell anaemia are examples.

Sex-linked inheritance is the inheritance of genes that are located on chromosomes based on gender. Since males only have one X chromosome, X-linked traits tend to be more prevalent. A male who inherits a recessive X linked gene will manifest the trait as he has no second X chromosome. Colour blindness and haemophilia are examples.

Scientists have been able to identify and study specific genes that are associated with different traits and disorders. This knowledge is important for understanding disease inheritance, genetic counselling and developing potential treatments and interventions.

Genetics is an exciting field that unravels mysteries such as genes, sex, and inheritance within families. Identification and characterization genes in an organism's DNA provide valuable insight into the molecular foundation of life and diversity of living organisms.

Understanding how genes affect our sex, and how traits are passed down through generations, not only enriches the knowledge we have of human biology, but it also has applications in many fields such as medicine, agriculture and evolutionary biology. Genetic research is advancing rapidly and opening new frontiers for knowledge. It could lead to breakthroughs in diagnosis, treatment and prevention of genetic disorders.

In exploring the complexities of genetics, we can gain a better understanding of the mechanisms that govern the biological makeup of our bodies and the interconnectedness between all the living organisms. Genetic research promises to unlock even more secrets about life and help us better understand ourselves as well as the world in which we live.

Chapter 10

Physical Sciences

Physical science is an area of science that focuses on the fundamental principles and concepts of physics and chemical. Physical science is the study of matter, energy, and their interactions in different forms. Physical science aims to discover the laws and forces that govern our physical world. This includes everything from subatomic particles up to the vastness the universe.

Chemistry

Chemistry is the branch of science that studies composition, structure, and properties of substances. Chemistry is a key part of understanding the world, from the everyday substances to the molecular reactions. This essay will explore the fascinating world of chemistry. We will cover topics like how chemists view things, the periodic tables and elements, changes in state of matter, mixtures, reaction, and properties of acids and bases.

Chemists use a systematic method to study matter, its transformations and their properties. To analyze and understand properties and behaviours of substances, they use experimentation and critical thinking. To gather information, chemists use physical and chemical properties such as colour and odour to observe substances.

Chemists can identify and quantify a substance's components using chemical analysis techniques such as spectroscopy or chromatography. Chemists experiment to determine the effect of conditions such as temperature and pressure on chemical reactions.

Understanding the Elements - The Periodic Chart

The periodic table is a grouping of substances that are arranged according to their atomic number, electron arrangement and chemical properties. The periodic table is used to organize and categorize the elements. The elements are arranged in groups of columns and periods of rows.

The periodic table helps chemists predict properties and understand relationships between elements. The similarity of electron configurations in elements within the same group can lead to similar chemical behaviours. The periodic table has distinct properties for transition metals, actinides, lanthanides and lanthanides.

Molecular Molecules Changing States - Physical and Chemical Movements

Solid, liquid and gas are all states of matter. Temperature and pressure variations can cause changes in the state of matter. Physical changes are alterations to the state or form of matter, without changing its chemical makeup. Examples include melting and freezing, boiling and condensation.

On the other hand, chemical changes involve rearrangement of the atoms in order to create new substances that have different chemical properties. Chemical reactions such as combustion and oxidation result in the breaking or formation of chemical bonds.

Boiling and freezing:

Both boiling and freezing are important physical processes that change the state of matter. When a liquid reaches the boiling point, it changes to a gas. Freezing occurs when a fluid reaches the freezing point. The boiling and freezing points of a substance depend on the chemical composition and surrounding pressure. These points provide information on the behaviour of substances, and they are used for a variety of applications such as cooking and preservation of food.

When elements get together, they form compounds, mixtures and reactions.

Compounds consist of a fixed number of elements combined chemically in a certain proportion. Compounds are made up of elements that are held together with chemical bonds. Compounds possess distinct properties that are different from those of their constituent elements.

The mixtures are a combination of two or three substances that have been physically mixed together, but are not chemically bound. Mixtures are homogeneous or heterogeneous. Each component in a mixture retains the original properties.

Chemical reactions are the result of the formation and breaking of chemical bonds. Reactants become products. The reactants are the initial substances and the products are those that are formed after a chemical reaction. These reactions are represented by chemical equations, which provide a concise description of the substances involved and their stoichiometry.

Covering All the Bases and the Acids

Acids and bases are important substances that have distinct chemical properties. Acids and bases are two types of substances with distinct chemical properties.

They are usually slippery and bitter to taste. They can neutralize acid and change the colour of indicators. Bases include ammonia and sodium hydroxide.

Acids have a sour flavor and can cause metals to corrode. Acids can change the color indicators, and they can react with bases to produce salts and water. Acids include hydrochloric (HCl), and acetic (CH3COOH) acids.

Chemistry is an exciting field that helps us understand the composition, properties and transformations in matter. We can better understand chemistry by exploring the way chemists view things, the periodic tables and elements, changes in state of matter, compounds and mixtures, reactions and properties of acids and bases.

It is essential to the advancement of science and technology, including drug development and materials syntheses as well as environmental studies and energy generation. It allows us to understand the substances that we come across in our everyday lives, such as the air we breathe and the food we eat.

We can learn and apply the principles of Chemistry to make better decisions, find innovative solutions for societal problems, and gain a deeper understanding of our world. Chemistry is a discipline that has paved the way for many discoveries and advances in various fields.

Physics

Physics is an area of science which aims to discover the principles that govern the behaviour of energy and matter. Physics is a broad field of study that includes a variety of phenomena from subatomic particles at the microscopic level to the vastness of the universe. This essay will cover a variety of aspects of physics including SI units, weight, mass, Newton's Laws of Motion, work, force and energy, power and sound waves.

Keep Track of SI Units:

The International System of Units provides a standard framework for measuring physical quantity. The SI includes seven basic units: meter for length, kilogram for mass, second for time, ampere for electric current and kelvin for temperature.

SI prefixes, such as mega-, kilo, milli, and micro, are used to indicate quantities that are larger or smaller than a base unit. As an example, the kilometre (km), which represents 1,000 meters in length, is equal to one thousandth of a second.

Weight and mass:

In everyday language, weight and mass are used interchangeably, but they have different meanings in physics. The mass of an object is the amount of material it contains. It remains constant, regardless of gravitational fields. It is usually measured in kilograms.

Gravity is what exerts the weight on an object. The weight of an object depends on its mass and the strength or gravity field. The weight is measured in newtons, and can vary depending on location.

Newton's Laws of Motion:

Newton's laws describe the motion of objects. Newton's Law of Motion, also called the Law of Inertia, explains how objects move. According to the law, if an object is not in contact with a force external, it will stay at rest. If an object is moving, it will maintain its speed unless it comes into contact with an external force. Newton's Second Law of Motion relates an object's mass and acceleration to the force that is acting on it. It states that acceleration is directly proportional with the net force acting on an object and inversely correlated with its mass. Newton's Third Law of Motion says that every action has a reaction. When two objects collide their forces are equal in magnitude and have opposite directions.

Measurement of Work, Force Energy and Power

In physics, the concepts of force, power, work, and energy are all interconnected. The work is the product of force and displacement. It is measured in Joules (J). The force is a quantity vector that accelerates objects. It is measured in Newtons (N). Energy is the ability to perform work or transfer heat. Energy can be found in a variety of forms including thermal, kinetic, and potential energy. The unit of energy is joules. Power is the rate of work or energy transfer. Power is the ratio between the work or energy done and the time taken. The power is measured in Watts (W).

Understanding Sound Waves:

Sound waves are mechanical waveforms that propagate in a medium such as air or water. The characteristics of sound waves are their frequency, wavelengths, amplitudes, and speeds. The frequency is the number of oscillations of a sound per unit of measurement and is measured by hertz. The wavelength is the distance between consecutive points of a soundwave that are in phase. For example, two crests and two troughs. It is measured in metres (m). The maximum displacement of particles from equilibrium is called the amplitude. It is the measure of loudness or intensity. The speed of sound is the rate that sound waves travel through a medium. The speed of sound depends on the properties and temperature of the medium. It is 343 meters per seconds (m/s), in dry air.

The Rainbow of Frequencies in the Electromagnetic Spectrum

The electromagnetic spectrum includes a wide range of electromagnetic waves. These include radio waves and microwaves. It also includes visible light, ultraviolet radiations, X-rays and gamma radiations. Waves are classified by their frequency, wavelength and energy.

The radio waves have the shortest wavelengths and the lowest frequency, while the gamma-rays have highest frequencies and the longest wavelengths.

Visible light is a range of wavelengths and frequencies that can be seen by the human eye. The rainbow colors, from red to purple, are included.

The interaction of electromagnetic waves with matter is unique. They can be refracted or diffracted by the medium in which they are encountered.

Heat Things Up With Energy

Heat is an energy form that moves from areas of high temperature to lower temperatures. Heat can be transmitted through conduction or convection. Conduction is the direct transfer of heat between particles within a solid. Convection is the transfer of heat by moving fluids such as air and liquids. Radiation is the transfer of heat via electromagnetic waves such as infrared. Thermal energy is internal energy that a system generates due to its random particle motion. It is influenced by factors like temperature, mass and specific heat capacity.

Magnetism and Polarization

The magnetism of certain materials is their ability to attract or repel materials in the presence of magnetic field. Magnetism is the behavior and alignment of microscopic magnetic fields within a material. North and South poles are the two poles of magnets. Poles that are similar repel each other while poles opposite attract. Magnetic fields exert force on magnetic materials as well as moving charged particles.

The polarization of electromagnetic waves is the alignment in a particular direction. This is done by blocking or filtering waves that vibrate in certain directions.

Physics is a broad field that encompasses many phenomena from the micro to the macro. We gain a better understanding of physical principles by studying SI units, weight, mass, Newton's Laws of Motion, work, force and energy, sound waves, electromagnetic spectrum, heating and magnetism.

The tools and knowledge of physics allow us to understand and predict the behaviour and properties of matter and energies. This allows us to create new technologies, solve real-world problems, and explore the mysteries and wonders of the universe. Physics is a discipline which continuously pushes boundaries in human knowledge, and leads to innovation across many fields. From engineering and medicine to renewable energy and astrophysics. Unlocking the secrets to physics allows us to gain a better understanding of our world and ourselves.

SECTION 5
TECHNICAL SKILLS

Chapter 11

Auto Information

This chapter will examine the different components that comprise a vehicle's bonnet. These components all work together to maintain the functionality and smooth operation of an automobile. Each part, from the chassis and frame to brake system plays a vital role in the overall safety and performance of the vehicle. Understanding these components will help drivers to better understand how their vehicle operates and make informed decisions about maintenance and repairs.

Chassis and Frame: Holding It All Together

The chassis and frame are integral components of the vehicle that provide structural integrity and foundation for a safe and reliable operation. The chassis and frame work together to support and balance the vehicle's weight and the various components. This ensures stability and handling.

The chassis is the backbone of a vehicle. It comprises the major structural elements that support and connect the various systems. The chassis is usually made from strong materials like steel or aluminum, to withstand forces and stresses during vehicle operation. The chassis is a stable platform that allows the body, engine and suspension to be attached.

The frame is the structure that the body and the mechanical components are attached to. The frame is used to support the weight of the vehicle and distribute it evenly over the chassis. To achieve strength and rigidity, the frame is usually constructed with a combination box sections, beams and cross members.

In vehicle engineering, the design and construction are critical. The chassis and frame must be able to withstand a variety of forces such as those encountered during acceleration, brakes, corners, and uneven roads. The chassis and frame provide structural rigidity that enhances the handling and stability of the vehicle.

The chassis and frame are also important for crash protection. Crumple zones are built into modern vehicles, and are areas of the frame and chassis that are engineered to absorb energy and dissipate it in the event of an accident. It helps protect passengers by reducing the force transfer to the passenger compartment.

During the manufacturing process, chassis and frames are subjected to strict testing and quality assurance measures. The engineers and technicians make sure that the chassis and frame meet safety and performance standards. This involves stress testing, durability assessment, and performance evaluation under different operating conditions.

The longevity and safety a vehicle depends on the proper maintenance and care given to the chassis and frame. It is important to conduct regular inspections of the vehicle for signs such as damage, corrosion or wear. To prevent further damage and safety hazards, any issues must be addressed immediately. Regular maintenance includes lubricating moving parts, checking alignment and fixing any steering or suspension issues.

The chassis and frame are the foundation for a vehicle. They provide structural support, stability and crash protection. These are essential components that need to be manufactured and designed to meet the demands of a vehicle. It is important to understand the role of the chassis and the frame in maintaining vehicle performance and structural integrity.

Different strokes for the engine

It is clear that the engine is at the core of every vehicle. The engine is responsible for converting potential energy in fuel to mechanical power that propels the vehicle. Understanding engines and their components and types can provide valuable insights into the complex process of power production and vehicle performance.

The most common engines are gasoline, diesel and electric. Spark-ignition gasoline engines, or gasoline engines, use a spark plug in the cylinders to ignite an air/gas mixture. The combustion process produces expanding gases which push the pistons downward, converting energy into rotational motion.

Diesel engines are compression-ignition machines. The engines compress air only in the cylinders and raise the temperature until diesel fuel is injected into combustion chambers. The combustion forces the piston to descend, converting energy into mechanical force. Diesel engines are known to be efficient and have high torque output.

The environmental friendliness of electric engines and the advancements in battery technologies are driving their popularity. Electric engines are powered by batteries that store electrical energy.

The motors then convert the electrical energy to rotational motion. They are suitable for hybrid and electric vehicles as they produce no emissions and have instant torque.

There are components that are used in all engines. The combustion takes place in the cylinders. Vehicles may have anywhere between four and twelve cylinders. High-performance cars can even have more. Each cylinder has a piston which moves up and down. The fuel-air mix ignites and creates pressure, which forces the piston downward, converting linear motion to rotational motion via the crankshaft.

The crankshaft forms the core of an engine. The crankshaft converts the reciprocating piston motion into rotary motion that drives the transmission, and eventually turns the wheels of the vehicle. Connecting rods connect the crankshaft to the pistons. This allows the pistons' linear motion to be translated into the rotating motion.

Camshafts, valves and intake and exhaust systems are also essential components. Fuel injection and carburetion systems are also important. The camshaft controls valves that regulate the flow and exhaust of fuel and air into the cylinders. The intake system makes sure that the engine gets a good air-fuel mix, and the exhaust system removes the combustion products.

Modern engines use fuel injection systems, which are controlled electronically. Fuel injectors spray the fuel precisely into cylinders. It allows for improved fuel efficiency, power output and emissions control. The older carburettors that were used to mix fuel with air prior to entering the cylinders perform a similar task.

Understanding engine components and their workings goes beyond simple curiosity. This knowledge allows individuals to make more informed decisions regarding vehicle maintenance, performance upgrades and troubleshooting. Knowing the engine types and characteristics can help you choose the best vehicle to meet your needs. It also provides a solid foundation for those who wish to pursue careers in automotive technologies, where knowledge of engine diagnosis and repair is essential.

The engine converts fuel into mechanical energy. The different engine types such as gasoline, diesel and electric engines all have unique performance characteristics. Understanding components like cylinders and pistons can provide valuable insight into the power-generation process. It is useful not only for car owners, but also for those who are interested in careers in the auto industry.

Electrical and Ignition Systems

The electrical system of a car plays an important role in the vehicle's safety, comfort and functionality. It powers and controls various components. The electrical system is composed of a variety of subsystems and components, each with a distinct purpose for providing power and ensuring proper operation.

The battery is a key component of the electrical system. The battery is a device that stores electrical energy. It also provides the initial power required to start the motor. The battery also provides power for electrical accessories, when the engine isn't running or the demand exceeds the charging system. Modern vehicles have rechargeable batteries that are designed to handle the demands of electrical systems.

Alternator is another important component of the electrical systems. The alternator is a generator that converts mechanical energy from the motor into electrical energy. It not only provides power to the electrical systems, but also charges the battery when the engine is running. The alternator ensures that the vehicle's electrical systems are kept running properly by supplying a constant supply of power.

The ignition system starts the engine and ensures proper combustion. The ignition system is composed of several components which work together to ignite the fuel-air mix in the cylinders by creating a spark. The ignition coil produces a high-voltage signal that is transmitted to the spark plugs via the ignition cables and wires. The spark plugs create the spark required to ignite combustion in each of the cylinders, which results in the engine running smoothly.

The electrical system is responsible for many vehicle components, including the starter motor. Electrical power is used to illuminate and improve visibility for lighting systems such as headlights and taillights. Electrical power is also used to power the radio, infotainment, climate control, power windows and other conveniences.

The electrical system in modern vehicles is tightly integrated with the other vehicle systems. It communicates with engine control units (ECUs) to monitor and regulate engine performance, emission, and other vital functions. It communicates with the BCM to control various electrical functions, such as window and door locks. The electrical system also interacts with other safety systems such as stability control and airbags, which contributes to vehicle safety.

It is important to maintain and care for the electrical system in order to achieve its best performance. The battery should be inspected for damage or corrosion, the alternator tested for charging efficiency, and the spark plugs checked for condition. Preventive maintenance and routine inspections can identify issues before they cause electrical system malfunctions.

Knowing the electrical system in a vehicle allows individuals to diagnose electrical problems and troubleshoot them. This knowledge allows them to make more informed decisions about replacing parts, upgrading electrical accessories or making electrical modifications. For those pursuing careers as automotive technicians, an understanding of the electrical systems is crucial for diagnosing, repairing, and maintaining vehicles.

The electrical system in a car is responsible for powering and controlling various components. This enhances safety, comfort and functionality. The alternator and battery provide the electrical power while the ignition system ensures the proper combustion of the engine. Understanding the electrical system allows vehicle owners to troubleshoot and maintain electrical issues and supports individuals who are pursuing careers within the automotive industry.

Avoiding a meltdown with cooling systems

A vehicle's cooling system is a vital component. It regulates the engine temperature and prevents it from overheating. The cooling system is essential for maintaining the optimal conditions of the engine, and to ensure its performance and longevity. It is important for car owners to understand the cooling system as this will allow them to maintain engine health, and avoid potential damage from overheating.

The cooling system's primary purpose is to dissipate excessive heat produced by the combustion process in the engine. Engines are operated at high temperatures. Without a cooling system the heat buildup could cause severe damage such as warped or blown cylinder head gaskets.

The cooling system is composed of several components that all work together to remove heat. The radiator is one of the most important components, as it is responsible for cooling down the heated coolant. The radiator, which is located usually at the front end of the car, consists of several small tubes and fins. The coolant is cooled by air flowing over the fins as it flows through the tubes.

The cooling system relies upon a water-pump to circulate coolant throughout the engine and radiator. Water pumps are usually driven by a chain connected to the crankshaft of the engine. The water pump creates pressure that circulates the coolant and ensures a constant flow of the fluid between the engine, radiator and the water pump.

The thermostat is an important part of the cooling system. It acts as a temperature-sensitive valve that regulates the flow of coolant based on the engine's operating temperature. The thermostat is closed when the engine's temperature is low, which allows it to heat up quickly. The thermostat opens once the engine reaches the operating temperature. Coolant can then flow through the radiator and maintain the engine temperature.

The mixture is designed to protect the engine against freezing and corrosion. It's also called coolant. It is more effective under extreme temperatures because it has a lower freezing point and a higher boiling temperature than water. Coolant is used to dissipate heat, but it also protects various cooling system components from corrosion and lubrication.

Understanding the cooling system will help you maintain engine health and prevent overheating. The cooling system components such as the hoses and belts or the radiator can be inspected regularly to identify signs of wear. Maintenance includes replacing and flushing the coolant according to the recommended intervals. This will ensure that it is effective and prevent contaminants from building up.

Proper driving habits, in addition to routine maintenance, can also contribute to the efficiency of the cooling system. By avoiding prolonged idling in hot weather and by not overloading your vehicle, you can reduce the heat stress that the engine experiences. The engine can be saved by monitoring the temperature gauge in the instrument panel, and responding quickly to any signs of overheating such as steam or strange smells.

It is important to understand the cooling system for those who drive in extreme weather conditions, or for those who tow or perform other tasks that require a lot of power. This puts additional pressure on the engine and makes efficient cooling more important to prevent overheating.

The cooling system is an important component of any vehicle. It regulates engine temperature, prevents overheating, and maintains optimal operating conditions. Radiator, water pump and thermostat work together to dissipate the heat and maintain the engine's performance and longevity. Understanding the cooling system allows vehicle owners to maintain their engine's health, avoid damage from overheating and react appropriately to any signs that indicate a cooling system malfunction.

Lubrication systems help keep the engine running smoothly.

It is important to reduce friction in the lubrication system of a car. This will ensure that engine components operate smoothly. It is crucial in extending engine life and ensuring optimum performance. Understanding the lubrication systems helps vehicle owners understand the importance of proper lubrication, and its impact on engine health.

The main purpose of the lubrication is to reduce friction and minimize metal-to-metal contact. The lubrication system reduces friction and wear, as well as excessive heat production that can damage the engine. The lubrication ensures that the engine's vital components such as the camshaft and crankshaft are properly lubricated for smooth operation.

Oil pump, oil filter and oil pan are all part of the lubrication systems. Oil is circulated throughout the engine by the oil pump. The oil pump draws oil from the pan, pressurizes it and distributes it to engine components via a network oil passages. The oil filter filters out impurities from the oil and prevents them from circulating and damaging the engine. Oil pans are used to collect and store oil when not in use.

Maintaining the health of your engine requires proper lubrication. Regular maintenance includes replacing oil filters and checking oil levels and quality. Lack of lubrication may lead to increased friction, wear and tear on the engine, and even failure.

Understanding the lubrication systems of their vehicles, vehicle owners will be able to appreciate the importance using recommended oil viscosity for their engine. The lubrication needs of different engines can vary. Using the wrong oil type could lead to poor performance, damage, and reduced lubrication. Follow the manufacturer's recommendations and guidelines for oil selection.

Exhaust systems: where byproducts are sent to die

The exhaust system of a car serves many important functions. It reduces noise and removes combustion by-products. The exhaust system is crucial to vehicle performance, the environment, and passenger comfort. Understanding the exhaust system allows drivers to understand the environmental impact of their vehicle emissions, and emphasizes the importance of emission controls.

The exhaust system's primary purpose is to reduce the noise generated by the engine when it burns. The exhaust system consists of components like the catalytic convertor, exhaust manifold and muffler. The exhaust manifold channels the exhaust gases of each cylinder into one pipe. The catalytic convertor is an important component which helps to reduce harmful emissions through the chemical conversion of toxic gases. The muffler reduces noise further by using sound-dampening material and internal chambers that dissipate sound waves.

The exhaust system also has another important function: the removal of emissions and combustion byproducts. Carbon monoxide and nitrogen oxides are produced by the combustion process of an engine. Unburned hydrocarbons are also released. The exhaust system makes sure that these gases can be safely released from the vehicle into the air. Modern vehicles have emission control systems that include oxygen sensors and exhaust gases recirculation valves (EGR). These help reduce harmful emissions.

By understanding the exhaust system, drivers can recognize the impact of their vehicle's emission on the environment and the air quality. Drivers can minimize their vehicle's carbon footprint by ensuring that the exhaust system is working properly. This includes regular maintenance and inspection.

A properly functioning exhaust system is also crucial to passenger safety and comfort. Damaged or malfunctioning exhaust systems can cause increased noise, unpleasant odors and health risks due to harmful gases being released into the vehicle's cabin. It is important to conduct regular inspections of the exhaust system. This includes checking for any leaks, corrosion or damaged components.

The lubrication and exhaust systems are essential components for the engine and emission control of a car. The lubrication ensures that engine components operate smoothly by reducing friction, preventing excessive wear, and generating heat. Understanding the lubrication systems allows vehicle owners to understand the importance of proper maintenance and lubrication for engine health. The exhaust system is responsible for reducing noise and removing combustion by-products. It also plays a vital role in ensuring passenger comfort and minimizing environmental impact. Understanding the exhaust system helps drivers understand the importance of emission control, and the necessity for proper maintenance in order to ensure eco-friendly and safe driving.

Filtering out Pollutants: Emissions Control Systems

Emissions control systems are designed to reduce harmful pollutants that vehicles emit. These systems use components like the oxygen sensor and catalytic convertor to transform harmful gases into less dangerous substances. Understanding emissions-control system is important for environmental sustainability and compliance with emission regulations.

Take a spin with Drive Systems

The drive system is responsible for transferring power from the motor to the wheels. This allows the vehicle to move. The drive system includes components like the transmission, driveshaft and differential. Different drive systems such as rear-wheel, front-wheel, and all-wheel offer different advantages and characteristics.

Suspension Systems and Steering: Keep it on the road

The suspension system is designed to absorb shocks and vibrations. This allows for a comfortable and smooth ride. The suspension system includes springs, shock-absorbers, and control arm components. The steering system is used to steer the vehicle. It includes the steering wheel and steering column.

Brake System: Pulling out all the stops

It is the brake system that slows down and stops the vehicle. The brake system includes brake pads, brake rotors and calipers. It is important to understand the brake system, and how it should be maintained.

Under the hood, all of the components work in harmony to give a vehicle power, control and safety. Each component, from the chassis and frame to brake system, plays a vital role in the overall performance and functionality of an automobile. Knowing these components will help drivers make informed decisions about maintenance, repairs and upgrades. This will ensure the longevity and performance of their vehicle. Understanding the intricate components under the hood allows drivers to appreciate the engineering marvel of an automobile, and encourages safe and responsible driving.

Chapter 12

Shop Information

In a shop, measurement tools are essential for ensuring precise and accurate dimensions. Craftsmen use tools like tape measures, rulers and calipers for accurate measurements. These tools allow craftsmen to align work correctly and ensure their projects meet required specifications. Craftsmen can achieve high-quality results by using measurement tools.

Many shop projects require that alignment and evenness be maintained in addition to the measurements. Laser levels and spirit level are essential tools for horizontal or vertical alignment. These tools allow craftsmen to ensure their work is aligned and level, which is important for many applications such as installing shelving, aligning windows and doors, or building frameworks. Squaring tools such as combination and framing squares can be used to ensure precise cuts and right angles. Craftsmen can achieve the level of accuracy and precision they desire by using levelling and square tools.

For tasks that require force or impact, striking tools are essential. In the workshop, striking tools such as hammers, rubber mallets and mallets are used. These tools are used by craftsmen to drive nail into materials, form or shape materials, or to provide the necessary impact in order to loosen or remove components. The type of tool to use depends on the task, since different materials or applications require varying levels of force and impact. Craftsmen can minimize the risk of injury or damage by selecting the right tool and using it.

Fastening tools ensure that materials are securely joined together. Fastening tools are commonly used in shops. Craftsmen can use these tools to assemble and disassemble components, tighten fasteners or loosen them, or create strong connections for various applications. Fastening tools can be used to ensure the durability and stability of the final product, whether it is assembling furniture or machinery, or building structures. Craftsmen can improve the quality and durability their projects by using the correct fastening techniques and tools.

When it comes to joining metals, welding and soldering equipment are essential in the workshop. Soldering irons and solder are used in delicate electronic applications. These tools enable craftsmen to make precise and secure connections, which ensure the correct functioning of electronic devices and circuits. In contrast, welding machines and electrodes work best for heavy-duty applications that require strong and permanent connections. Welding allows craftsmen to join metal pieces, creating joints that are robust and can withstand heavy loads and harsh environments. Craftsmen can choose the best technique for each project by understanding the various soldering and weld tools and their applications.

Cutting tools are essential for shaping and dividing different materials. Cutting tools are used in many shops. They come in a variety of types and sizes to fit different needs. Circular saws and handsaws allow for straight, curved or intricate cuts to be made in wood, metal, plastic or other materials. Each saw type has its advantages, and each is made for specific materials or cutting needs. Cutting tools such as utility knives and scissor are also used to perform finer, more precise tasks. These tools allow craftsmen to make clean, accurate cuts whether trimming material, cutting fabric or performing detailed work. Craftsmen can cut materials and fabric with precision and efficiency by using the right cutting tools and techniques.

Tools such as gouging, drilling, and punching are necessary to create holes, remove material, or make indentations on various projects. Tools like drill bits, hole saws and punches are some of the tools that allow craftsmen to accomplish these tasks efficiently. There are many different sizes and types of drill bits to suit different materials and hole size. Hole saws have been designed to cut larger holes into materials such as wood and metal. Punches can be used to make marks or indentations on materials. They are often used as a starting point for drilling, or to align the material. Chisels can be used to shape, remove material, or carve. Craftsmen can shape and modify materials using drilling, punching and gouging instruments.

Soldering tools, welding tools, cutting tools and drilling, punching and gouging are all essential tools for a well-equipped workshop. These tools allow craftsmen to join materials, create openings and indentations, or shape metals as needed for their projects. Craftsmen can ensure precise and reliable results by understanding the capabilities of these tools and their applications. For optimal performance, it is important to maintain these tools properly, by sharpening drills or replacing worn out saw blades. Craftsmen can achieve amazing results in their shops with the right tools and methods.

Finishing tools are essential for achieving the smoothness and aesthetics desired in a project. These tools are used for improving the appearance, the edges and the surface of the material being worked on. Sandpaper is one of the most popular finishing tools. It comes in different grits, and it can be used to smooth rough surfaces, remove imperfections and create a uniform surface. Also, files and rasps can be used to refine and shape edges, contours and curves. These tools enable artisans to achieve the desired smoothness and accuracy. They ensure a polished and professional finish for their projects. Craftsmen can turn raw materials into visually pleasing creations by using finishing tools.

Clamping is essential to secure materials during different shop operations. These tools offer stability, precision and control. They ensure that the workpiece is firmly fixed while drilling, gluing or cutting components. In the workshop, clamping tools such as vices and Clamps are commonly used. Clamps are available in many types such as bar clamps and spring clamps. They can be used to hold materials of different sizes and shapes. Vices are used to clamp heavier or larger items. These clamping tools enable craftsmen to work accurately and with confidence, as they prevent materials from shifting or moving. Clamping tools are essential for maintaining precision and integrity of workpieces, whether it is ensuring straight cut, drilling precise holes or achieving seamless adhesive joints.

A comprehensive collection of these tools allows for craftsmen to complete a variety of tasks with efficiency and precision. Understanding the functions of each tool and how to use them is crucial for high-quality results on any project. For optimal performance and longevity, it is important to maintain tools properly, which includes regular cleaning, sharpening and calibrating. Maintaining tools in good shape ensures reliability and increases their lifespan. This ultimately contributes to the efficiency and success of your shop. The tools of the trade are essential in the shop to accomplish tasks efficiently and achieve quality craftsmanship. Each tool has its own function and is essential to various shop operations. Understanding their purpose and how to use them properly allows craftsmen to achieve success and satisfaction in their projects by working with confidence, precision and efficiency.

Sticking Materials Together with Fasteners

They are essential in securing materials. Fasteners provide the strength and stability needed to secure different materials or components. Fasteners are available in many types and shapes, each one suited to specific requirements and applications. Here are some of the most common fasteners found in shops:

Nails: Nails have been around for centuries and are the simplest of all fasteners. They are usually made from metal with a pointed tip. The nails are driven in with a nail gun or hammer. In woodworking, they are used to join wood pieces together. Nails are strong and reliable, especially if driven through thicker sections of material.

Screws and Bolts - Screws and bolts provide strong, adjustable connections. The bolts and screws consist of a shaft with a threaded head on one end. Bolts are fastened with nuts and bolts have threads which cut into the material. Bolts and screws are used for a variety of purposes, such as furniture assembly, machinery, construction and more. The bolts and screws are removable, allowing for disassembly or adjustments when necessary.

Nuts, washers, and screws: These are commonly used with bolts and screws. Nuts are fasteners with threads that are attached to the threaded shafts of bolts. They create a thread that matches the bolt to provide a strong hold. They are thin, flat plates that go between the nut of the fastening material and the washer. They spread the load, provide stability and prevent the nut from losing.

Rivets - Rivets can be used to connect materials by creating a solid, strong connection. The head is attached to a cylindrical shaft. The shaft of the rivet is "peened" or deformed to ensure the connection. Rivets can be inserted in pre-drilled holes on the materials that will be connected. In metalworking, aircraft building, and automotive applications, rivets are frequently used. They are a durable and reliable connection, resistant to movement and vibration.

It is important to understand the various types of fasteners, their uses and how they can be used in a particular project. When choosing the correct fastener, you should consider factors such as the type of material to be fastened, required strength, and ease of disassembly. Craftsmen can use the right fasteners to ensure their creations will be held securely and reliably together. This contributes to the strength and durability the final product.

Chapter 13

Mechanical Comprehension

A-mass-ing Knowledge on Matter, Weight, Density, and Relativity

The understanding of weight, density and matter is based on forces. Forces are interactions between objects that can alter their state, shape or motion. You can learn more by exploring the fundamentals such as action and reaction, pressure, and equilibrium, and different types of forces.

Newton's Third Law of Motion states: "Every action has an opposite reaction." This principle explains the nature and interaction of forces within the physical universe. If one object exerts a force on another object, the second object will react with a force equal in magnitude and opposite in direction. This concept is obvious in many situations. This is evident when we push against a brick wall. The wall is unable to move through us. This principle is essential to understand vehicle propulsion. The force the vehicle exerts must be balanced with the reaction force which pushes it forward in the opposite direction. Understanding the fundamental principles of action, response and force will give us a better understanding of the relationship between forces.

Equilibrium is the state of equilibrium where forces are balanced and result in a stable situation. Both the net force and the torque acting on an object are zero in equilibrium. This means that torques and forces in opposite directions cancel each other out, leading to constant velocity or rest. Understanding equilibrium is crucial for analyzing and predicating the behaviour of objects when they are subjected to different forces. A stable object can move or maintain its position without any acceleration. The force of gravity pushing a book down on a desk equals the force of the table pushing the book up. Equilibrium can be used to determine the conditions needed for different scenarios. Designing structures that are resistant to external forces or ensuring mechanical system stability, for example.

The force is exerted by the area under pressure. It is the ratio between the force perpendicularly acting on a surface and the area that it is distributed over. The role of pressure is crucial in many aspects of daily life, including the hydraulic system and fluid dynamics. Understanding pressure allows us to understand how forces are distributed over an area, and how they impact objects that come into contact.

There are many types of forces. Some of the common forces are gravitational force and electromagnetic force. Gravitational forces is what attracts objects of mass. The electromagnetic force includes both magnetic and electric interactions. Normal force is the force that a surface exerts to support an object. The frictional force is the resistance to motion between objects that are in contact. When a cable, string, or rope is tightened, tension force is produced. A force that is applied by an outside agent. Understanding the types of force and their influence on objects allows us to analyze and predict how they will behave in different situations.

Understanding the forces that are at work in matter, mass, density and relativity will allow us to dig deeper into the principles that govern each of these concepts. Understanding the fundamentals of action and reactions, equilibrium, force, and other types of forces will help you better understand the behaviour and interactions of physical objects. We can gain a better understanding of the complex mechanisms that govern our understanding of weight, density and relativity by expanding our knowledge of these topics.

You Call That Work?

Contrasting the Difference Between Potential and Kinetic Energy

The difference between kinetic and potential energy is important in the world of physics. The potential energy of an object is determined by its position or condition. Potential energy is energy stored that can be converted to other forms of power. Potential energy includes gravitational energy, which is based on the height of an object above the ground. Another example is elastic potential energy which is stored by objects such as springs and stretched rubber bands. Kinetic energy, on the other hand is the energy that an object has when it's in motion. Both the mass and velocity of an object are important. The object moves when potential energy is converted to kinetic energy. Understanding the difference between potential and kinetic energies allows us to understand the interplay of stored and active energy in various systems.

Applying force to cause things to happen: In physics, force is a concept that describes a push or pull an object experience as a result of an interaction with another. A force can change the motion or shape of an object when it is applied. Forces are exerted by direct contact (such as pushing a cart) or fields such as electromagnetic or gravitational forces. We can change the shape of an object, their speed, or their state by exerting force. Understanding force principles allows us to predict and analyze how objects will react to external influences, and how we can manipulate their behaviour.

Resistance is the force or obstacle that opposes the movement or the change in the motion of an item. We often have to overcome resistance when we are trying to move or change an object's state. The resistance can take many forms such as friction, air pressure, or other objects. To overcome resistance, you must apply enough force to surpass and counteract the opposing forces. Understanding the nature of resistance, its effects and how to overcome them, will help us design more efficient systems and reduce energy loss. We can also optimize the performance of mechanical and physical processes.

Work is the key to powering: In physics work is defined by the transfer of energy when an object is moved in the direction that the force is moving. Calculating work is done by multiplying force magnitude by distance. Energy is transferred by this method. Work is the act of exerting force over a long distance. This can lead to energy being transformed from one form into another. Understanding the concept work allows us to analyze and quantify the energy transformations and transfers that take place in different tasks and processes. This allows us to assess the efficiency of systems and the amount of energy produced or required, as well as make informed decisions about energy use and conservation.

We can learn more about the basic concepts that govern the physical universe by exploring the differences between potential and kinetic energies, the principles of force and the recognition and overcoming of resistance. These concepts are crucial for understanding the dynamics of motion and energy. They also provide a basis for exploration and application to fields like engineering, mechanics and everyday life.

Working with Machines

Use levers to your advantage. Levers are simple machines consisting of a rigid beam or bar that pivots about a fixed point known as a fulcrum. You can increase the force of a lever by applying force to one end. The three main types of levers are based on their relative positions, their effort (the force they apply), and their load (the resistance that is being moved). Understanding the lever's principles allows us to maximize our force to accomplish our tasks and amplify it.

The ramping up of the inclined plane. An inclined plane is an inclined surface which allows you to exert a lesser force over a longer distance in order to move a heavy object from one level to another. Ramps, stairs and even road inclines are examples of inclined planes. An inclined plane can reduce the force needed to overcome gravity and move heavy objects by increasing the distance at which the force is applied. Understanding the mechanical advantage of inclined planes allows us to use these structures efficiently in order to reduce effort, and accomplish tasks which would otherwise be difficult.

Gears and pulleys can help you reduce your effort. They are devices that multiply and transmit force, or change direction. The wheel has a grooved edge and the rope or cable runs along it. The pulleys are used to lift heavy objects with less effort. They do this by spreading the force across multiple ropes, and changing the direction. Gears are teethed wheels which mesh to transmit motion and power. The gears allow us to convert speed and torque, increasing or decreasing the speed or force of a rotating motion. Understanding the principles behind pulleys, gears and other mechanical devices allows us to manipulate force and movement in various applications.

Wheels and axles multiply your efforts: A wheel and axle machine is made up of a wheel that has a large diameter (the wheel), and a small cylindrical object that rotates along with it (the axle). You can increase the rotational force at the axle by applying more force to the wheel. Wheel and axle arrangements allow for multiplication of forces, making it easier to rotate or move heavy objects. Understanding how the wheels and axles interact allows us to take advantage of their mechanical advantage and accomplish tasks with less effort.

Torque: A measure of force that can make an object rotate around an axis. The torque is affected by the force and distance from the rotation axis. We can increase torque by increasing the distance away from the axis or applying more force. It is important for tasks that require turning or rotating an object, like opening doors or tightening screw. Understanding torque allows us to select the correct tools and apply the proper amount of force in order to achieve rotational motion.

The vise is a device that holds objects in place firmly during different tasks. The vises consist of two jaws, one fixed, and the other movable. They can be tightened to hold an object. Vises give us a stable grip and allow us to work with objects without worrying about them slipping or moving. Knowing how to use vises correctly allows us to work safely and accurately. They provide the stability needed for tasks that need precision or force.

Hydraulic pressure. Hydraulic systems use the principle of Pascal's law which states that pressure applied to fluids in closed systems is distributed equally in all directions. This principle is used by hydraulic systems to transmit and generate force using fluids such as water or oil, along with a series interconnected tubes and pistons. Hydraulic systems generate greater force by applying a smaller force to a smaller piston. This provides a mechanical advantage. Understanding hydraulic systems enables us to harness fluid pressure in order to lift heavy objects, operate machinery and perform various tasks with precision and ease.

Understanding and using the principles of machines, such as levers and inclined planes and pulleys and wheels and axles and hydraulic systems can enhance our efficiency and work capability. These machines enable us to reduce effort, manipulate movement, and secure objects. This allows us to complete a wide range of tasks more efficiently and effectively. By incorporating these principles into our daily work, we can optimize productivity, increase safety and achieve the desired outcome more efficiently.

Chapter 14

Electronics Information

The Secrets of Electricity

Subatomic particles: Atoms are composed of protons, neutrons and a nucleus, which is surrounded by electrons orbiting around it. These electrons orbit the nucleus at specific energy levels, or shells. Electrons can be shared or transferred between atoms when they interact, causing an imbalance of their charges. The movement of electrons is what causes electrical phenomena. Scientists have gained a deeper understanding of electricity by studying electron behaviour and interactions in atoms.

The good conductor medal is awarded to materials that are both conductors and insulators. Both play an important role in the flow electricity. Materials that are conductors have a low resistance to electric current flow. These materials allow electrons to freely move within their atomic structures, which facilitates the flow of electric current. Due to their mobility, metals such as aluminium and copper are excellent conductors. Insulators are materials with a high resistance to electric current flow. They stop the flow of electricity and the movement of electrons. Insulators are often made of rubber, plastic and glass. It is important to understand the properties of conductors, insulators and other materials when designing electrical systems. Engineers can choose materials that are conductive or insulative based on the properties they possess. This will ensure a safe and efficient flow of electricity for various applications.

Electrical current is the flow of charge through a conductor. It is the movement electrons from a higher electric potential area to a lower electric potential area. The units of measurement for current are called amperes, named after French physicist Andre-Marie Ampere. Electric current is available in two different forms: direct (DC) or alternating (AC). Direct current is a flow of electrons that moves continuously in one direction. Batteries and electronic devices commonly use direct current. Alternating current, on the other hand, changes direction periodically, oscillating. In most commercial and household electrical systems, alternating current is used. Understanding electric current's direction, magnitude and frequency is essential to harnessing the power of this electrical system.

Put the brakes on the electrical flow. Resistance is the property of materials which hinders or impedes electrical flow. The interaction between electrons, atoms, or molecules in a material is what causes it. The resistance is affected by a number of factors, including the composition, temperature and physical dimensions of a material. Resistors are components that introduce a certain amount of resistance to a circuit. They are used in electronic circuits to regulate voltage, control current flow, and limit current levels. Understanding resistance is important for designing circuits that have desired current levels. It also helps to prevent overheating and ensure the efficient and safe operation of electrical devices.

Measuring voltage. Voltage is also known as the electric potential difference. It is a measurement of electrical potential energy per charge unit between two points within an electrical circuit. It is the force that causes electrons to flow. Voltage is measured by volts, named after Italian physicist Alessandro Volta. Voltage is generated in many ways, including by batteries, power supplies or generators. The voltage measurement allows us to measure the potential difference in different points of a circuit. This information is valuable for determining the amount of electrical energy that can be used to power devices and perform work. Understanding voltage is important for circuit design and power supply requirements. It also ensures that electrical devices are compatible with available voltage levels.

Circuits are closed loops or pathways through which current can be carried. It is made up of components that are interconnected and perform specific tasks. A circuit usually includes an electrical source (such as a power supply or battery), conductors that carry current, and other components such as resistors, switch, and loads. The conductors are the conduits for electrons to flow, and the components change the behaviour of currents based on the characteristics of their parts. Understanding circuits requires analyzing current flow, voltage drops between components, and interplay of different circuit elements. This allows us to design circuits that meet specific requirements, to calculate current and voltage, to predict circuit behaviour and to diagnose and troubleshoot any circuit malfunctions. Engineers and technicians who understand the principles of circuits can design functional electrical systems that are efficient for many applications.

Measure power: Power is the rate of electrical energy consumption or production. It measures the amount of energy or work transferred per unit time. In electrical systems power is calculated by multiplying voltage (V), by current (I) that flows through the circuit. The power is measured in watts, where 1 watt equals 1 joule every second.

It is important to measure power in order to understand the efficiency and capacity of electrical devices. This allows us to evaluate the energy consumption of machines, appliances, and systems. We can then make informed decisions about energy conservation and usage. We can monitor power consumption to identify inefficiencies and implement measures that optimize energy usage and reduce costs and environmental impact.

Power measurements are also crucial for designing and sizing electric systems. Engineers can accurately determine the power requirements for various components and devices to ensure that their system can handle and operate at the required load. In addition to being important for safety, power measurements can also help reduce the risk of electrical hazards by preventing overloading circuits and electrical devices.

Electrical effects: Electricity is able to produce an amazing range of effects. Each effect has its own unique application and implications. Electric currents can produce a variety of effects when they flow through conductors, including light, sound, heat and electromagnetism.

The generation of light is one of the most common electrical effects. Incandescent lights, fluorescent tubes and LEDs are examples of devices that produce light. The electrical energy can be converted to visible light by passing an electric current though a filament or gas-filled tube.

Another common electrical effect is heat. Electric heaters and stoves transform electrical energy into thermal power, which is used to heat our homes, cook our food and power industrial processes. Understanding how heat is generated through electrical resistance helps us design heating systems that are efficient and to control the temperature.

Electrical effects also produce sound. Microphones transform sound waves into electrical signal, which is then amplified and played back through speakers. With this technology, we can communicate, listen to music and enjoy immersive audio in different settings.

Electromagnetism, a fundamental electrical phenomenon with many applications, is an important element of modern technology. Electric motors, transformers and generators are all based on electromagnetism. We can convert energy into motion, transmit electrical power over long distances, and create motion by harnessing the interaction of electricity and magnetism.

Understanding these electrical effects will allow us to harness electricity and manipulate it to suit our needs. It is responsible for technological advances in many industries including telecommunications and transportation. Exploring the principles that drive the production of electric effects can spark innovation, and open up new opportunities for improving lives and shaping our future.

By unravelling the secrets of electricity, we gain a deeper understanding of its fundamental principles and practical applications. From the behaviour of subatomic particles to the measurement of voltage, resistance, and power, electricity is a fascinating and powerful force that drives our modern world. Understanding and harnessing electricity allows us to light our homes, power our devices, communicate over long distances, and propel our technological advancements.

Alternating and Direct Current

Figuring out frequency: Frequency refers to the number of complete cycles or oscillations of an alternating current (AC) per second. It is measured in units called hertz (Hz). In AC power systems, the frequency is typically 50 Hz or 60 Hz, depending on the region. Understanding frequency is crucial for ensuring compatibility between electrical devices and power sources, as well as for designing and troubleshooting AC circuits.

Impedance: Join the resistance! Impedance is a measure of the opposition that an electrical circuit presents to the flow of alternating current. It includes both resistance and reactance (which is caused by inductance and capacitance). Impedance is measured in ohms (Ω) and affects the behavior of AC circuits. Understanding impedance allows us to calculate and optimize the performance of AC circuits, ensuring proper functioning and minimizing power losses.

Rectifying the situation: Going direct: Rectification is the process of converting alternating current (AC) into direct current (DC). This is achieved using diodes or rectifiers, which allow current to flow in one direction only. Rectification is necessary for many electronic devices that require a steady and unidirectional flow of current. Understanding rectification enables us to design power supplies, inverters, and other devices that convert AC to DC.

Turning up the transistor radio: Transistors are electronic components that amplify or switch electronic signals and currents. They are crucial for the operation of many electronic devices, including radios, televisions, and computers. Understanding the principles of transistor operation allows us to design and optimize electronic circuits, improving their efficiency and performance.

Decoding Electrical Circuit Codes

Wire Color Codes Used in the U.S: In the United States, electrical wires are often color-coded to indicate their specific functions and purposes. While color codes can vary between different regions and applications, some commonly used color codes include:

Black: Typically used for hot wires carrying current from the power source.

White or Gray: Used for neutral wires that complete the circuit and return current to the power source.

Green or Bare Copper: Used for grounding wires, which provide a path for electrical faults to dissipate safely.

Red or Blue: Occasionally used for hot wires in three-way switches or alternate systems.

Yellow: Used for wires that control switch circuits, such as three-way or four-way switches.

Orange: Used for control wires in some applications, such as irrigation systems or heating systems.

Brown: Occasionally used for high-voltage wires in certain applications.

Understanding wire color codes is essential for safe and proper electrical installations, as it allows electricians and technicians to identify and connect wires correctly. It helps ensure that electrical systems are wired in compliance with regulations and standards, reducing the risk of electrical hazards and improving overall system reliability.

Chapter 15

ASSEMBLING OBJECTS

Get the picture about Assembling objects:

Assembling involves assembling various parts to form a final structure or product. This requires planning, detail-oriented attention, and a thorough understanding of the way that different parts work and fit together. It is important to be able to visualize the process of assembly, whether it's for furniture, machinery or other complex items.

You get two types of questions for the price of one:

Connectors facilitate the joining of parts and components. Connectors come in many different forms, including screws, bolts and nuts, hinges, clips and fasteners. It is important to understand the different types of connectors and their functions in order to select the right ones for a specific assembly task. Different connectors provide different levels of stability, strength and disassembly. Their proper use ensures the integrity of the assembled object.

Shapes: Shapes are important in assembling objects, as they dictate how components will fit together. Objects come in many shapes including polygons and irregular shapes. Understanding the shapes and interactions of the parts is crucial for a successful assembly. Analyzing the dimensions, angles and contours of components is necessary to ensure proper alignment and fit. Understanding and interpreting shapes are crucial to determining the correct orientation and positioning of component, minimising errors and achieving an end product that works as intended.

Understanding connectors and shapes is important when assembling objects. It allows people to visualize assembly processes, understand instructions and diagrams, as well as execute necessary steps. This involves understanding how shapes interact with connectors to form a stable and functional structure. This knowledge allows individuals to confidently complete assembly tasks, troubleshoot problems, and produce well-assembled items with precision and efficiency.

Chapter 16

Practice ASVAB Exam

Question 1: Which of the following words is a synonym for "abundant"?

a) Scarce

b) Plentiful

c) Sparse

d) Limited

Answer: b) Plentiful

Explanation: "Abundant" means plentiful or in large quantities.

Question 2: Choose the word that is spelled correctly.

a) Accommodate

b) Acommodate

c) Accomodate

d) Acommadate

Answer: a) Accommodate

Explanation: The correct spelling is "accommodate," with two "c"s and two "m"s.

Question 3: Which word is the opposite of "courageous"?

a) Fearless

b) Brave

c) Timid

d) Strong

Answer: c) Timid

Explanation: "Timid" means lacking in courage or easily frightened.

Question 4:Identify the word that is a homonym for "sight."

a) Site

b) Cite

c) Seight

d) Siet

Answer: a) Site

Explanation: "Site" is a homonym for "sight" as they sound the same but have different meanings.

Question 5:Which word is a verb?

a) Beautiful

b) Happiness

c) Running

d) Tree

Answer: c) Running

Explanation: "Running" is a verb indicating an action.

Question 6:Choose the word that is a synonym for "exquisite."

a) Plain

b) Ugly

c) Delicate

d) Rough

Answer: c) Delicate

Explanation: "Exquisite" and "delicate" both mean highly pleasing or refined.

Question 7:Identify the word that is an adjective.

a) Carefully

b) Quickly

c) Smartly

d) Clever

Answer: d) Clever

Explanation: "Clever" describes a person or thing and is an adjective.

Question 8:Which word is an antonym for "victory"?

a) Success

b) Achievement

c) Defeat

d) Triumph

Answer: c) Defeat

Explanation: "Defeat" is the opposite of "victory," indicating loss or failure.

Question 9:Choose the word that is a synonym for "generous."

a) Greedy

b) Selfish

c) Kind

d) Stingy

Answer: c) Kind

Explanation: "Generous" and "kind" both describe a person's willingness to give or help.

Question 10:Identify the word that is spelled incorrectly.

a) Argument

b) Accomplish

c) Rhythm

d) Neccessary

Answer: d) Neccessary

Explanation: The correct spelling is "necessary" with one "c" and two "s"s.

Question 11:What is the main idea of the paragraph?

a) The importance of exercise for maintaining good health.

b) The benefits of a balanced diet and regular exercise.

c) The negative effects of a sedentary lifestyle.

d) The role of physical activity in preventing diseases.

Answer: b) The benefits of a balanced diet and regular exercise.

Explanation: The paragraph discusses the advantages of both a balanced diet and regular exercise for overall well-being.

Question 12:What does the author imply about the impact of technology on social interactions?

a) Technology has improved communication and made social interactions easier.

b) Technology has led to a decline in face-to-face interactions and weakened social connections.

c) Technology has had no significant effect on social interactions.

d) Technology has made social interactions more complicated and time-consuming.

Answer: b) Technology has led to a decline in face-to-face interactions and weakened social connections.

Explanation: The paragraph suggests that technology has had a negative impact on social interactions by reducing face-to-face interactions and weakening personal connections.

Question 13:According to the paragraph, why is reading important?

a) Reading expands vocabulary and enhances language skills.

b) Reading provides entertainment and relaxation.

c) Reading helps in acquiring knowledge and understanding the world.

d) Reading improves critical thinking and analytical abilities.

Answer: c) Reading helps in acquiring knowledge and understanding the world.

Explanation: The paragraph states that reading is important because it helps in gaining knowledge and understanding various aspects of the world.

Question 14:What is the author's opinion about the use of smartphones in classrooms?

a) Smartphones should be allowed in classrooms as they enhance learning.

b) Smartphones should be completely banned from classrooms.

c) Smartphones can be useful tools for learning if used appropriately.

d) Smartphones distract students and hinder their academic progress.

Answer: c) Smartphones can be useful tools for learning if used appropriately.

Explanation: The paragraph suggests that smartphones can be beneficial for learning if they are used appropriately in classrooms.

Question 15: What is the purpose of the paragraph?

a) To discuss the advantages and disadvantages of social media.

b) To provide tips for effective time management.

c) To explain the impact of climate change on the environment.

d) To highlight the benefits of volunteering in the community.

Answer: d) To highlight the benefits of volunteering in the community.

Explanation: The paragraph focuses on the advantages of volunteering and its positive impact on the community.

Question 16: According to the paragraph, what are the consequences of deforestation?

a) Loss of biodiversity and increased soil erosion.

b) Excessive rainfall and flooding.

c) Increased greenhouse gas emissions and climate change.

d) Reduced agricultural productivity and food scarcity.

Answer: a) Loss of biodiversity and increased soil erosion.

Explanation: The paragraph states that deforestation leads to the loss of biodiversity and increased soil erosion.

Question 17: What is the author's viewpoint regarding the use of renewable energy sources?

a) Renewable energy sources are expensive and inefficient.

b) Renewable energy sources are the future and should be prioritized.

c) Renewable energy sources have minimal impact on reducing carbon emissions.

d) Renewable energy sources are unreliable and cannot meet the energy demands.

Answer: b) Renewable energy sources are the future and should be prioritized.

Explanation: The paragraph suggests that renewable energy sources are important for the future and should be given priority.

Question 18: What is the paragraph mainly about?

a) The benefits of meditation for mental well-being.

b) The origins and history of yoga.

c) Different types of exercise for physical fitness.

d) The relationship between mindfulness and stress reduction.

Answer: a) The benefits of meditation for mental well-being.

Explanation: The paragraph discusses the advantages of meditation in promoting mental well-being.

Question 19: What is the author's purpose in providing statistical data in the paragraph?

a) To emphasize the seriousness of the issue being discussed.

b) To challenge the accuracy of the information presented.

c) To compare different perspectives on the topic.

d) To provide a historical background of the subject.

Answer: a) To emphasize the seriousness of the issue being discussed.

Explanation: The author uses statistical data to highlight the seriousness of the issue, supporting the argument being presented.

Question 20: What is the main point made in the paragraph?

a) The negative effects of excessive screen time on children's development.

b) The importance of parental involvement in children's education.

c) The benefits of extracurricular activities for children's social skills.

d) The impact of technology on the younger generation's communication skills.

Answer: c) The benefits of extracurricular activities for children's social skills.

Explanation: The paragraph discusses how extracurricular activities can enhance children's social skills.

Question 21: What is the value of $4 + 3 \times 2$?

a) 10

b) 14

c) 11

d) 8

Answer: c) 11

Explanation: According to the order of operations (PEMDAS/BODMAS), multiplication should be done before addition. Thus, 3 × 2 equals 6, and when added to 4, the total is 10.

Question 22: Solve the equation: 3x + 7 = 16.

    a) x = 3

    b) x = 4

    c) x = 5

    d) x = 6

Answer: b) x = 4

Explanation: To solve for x, subtract 7 from both sides of the equation, which gives 3x = 9. Then, divide both sides by 3 to isolate x, resulting in x = 4.

Question 23: What is the square root of 64?

    a) 6

    b) 8

    c) 4

    d) 10

Answer: b) 8

Explanation: The square root of 64 is the value that, when multiplied by itself, equals 64. In this case, the square root of 64 is 8.

Question 24: Simplify the expression: 5(2 + 3) - 4.

    a) 11

    b) 17

    c) 23

    d) 21

Answer: d) 21

Explanation: Begin by simplifying the expression inside the parentheses: 2 + 3 equals 5. Then, multiply 5 by 5 to get 25. Finally, subtract 4 from 25, resulting in 21.

Question 25: What is 35% of 80?

    a) 28

    b) 32

    c) 40

d) 56

Answer: a) 28

Explanation: To find 35% of 80, multiply 80 by 0.35. The result is 28.

Question 26: Find the average of 12, 15, 18, and 21.

    a) 16

    b) 17

    c) 18

    d) 19

Answer: c) 18

Explanation: To find the average, add up the numbers and divide the sum by the total count. In this case, $(12 + 15 + 18 + 21) \div 4$ equals 18.

Question 27: A rectangle has a length of 8 cm and a width of 5 cm. What is its perimeter?

    a) 13 cm

    b) 16 cm

    c) 26 cm

    d) 36 cm

Answer: b) 16 cm

Explanation: The formula for the perimeter of a rectangle is $P = 2(l + w)$, where l is the length and w is the width. Plugging in the values, we get $P = 2(8 + 5) = 2(13) = 26$ cm.

Question 28: If $6x + 4 = 22$, what is the value of x?

    a) 2

    b) 3

    c) 4

    d) 5

Answer: d) 5

Explanation: To solve for x, subtract 4 from both sides of the equation to get $6x = 18$. Then, divide both sides by 6, which gives $x = 3$.

Question 29: What is the value of $3^2 - 2^3$?

    a) -1

b) 1

c) 5

d) 7

Answer: a) -1

Explanation: 3² equals 9, and 2³ equals 8. Subtracting 8 from 9 gives -1.

Question 30: Simplify the expression: $4 + 2^2 \times 3$.

a) 16

b) 20

c) 24

d) 28

Answer: c) 24

Explanation: According to the order of operations, first square 2 to get 4. Then, multiply 4 by 3 to get 12. Finally, add 4 and 12, resulting in 16.

Question 31: Simplify the expression: $3x - 2(4x + 5)$.

a) -5x - 10

b) -5x + 10

c) -2x - 10

d) -2x + 10

Answer: c) -2x - 10

Explanation: Distribute the -2 to the terms inside the parentheses: $-2(4x + 5) = -8x - 10$. Then, combine like terms: $3x - 8x - 10 = -5x - 10$.

Question 32: What is the solution to the equation $2(x - 4) = 6x + 3$?

a) x = -3/8

b) x = -3/4

c) x = 3/8

d) x = 3/4

Answer: b) x = -3/4

Explanation: Begin by distributing 2 to the terms inside the parentheses: $2x - 8 = 6x + 3$. Next, subtract 2x from both sides: $-8 = 4x + 3$. Then, subtract 3 from both sides: $-11 = 4x$. Finally, divide both sides by 4 to solve for x: $x = -11/4 = -3/4$.

Question 33: Which of the following is equivalent to the expression $4(2x - 3) - 5(6 - x)$?

    a) $2x - 37$

    b) $7x - 27$

    c) $13x - 12$

    d) $24x - 17$

Answer: b) $7x - 27$

Explanation: Distribute 4 and -5 to the terms inside the parentheses: $8x - 12 - 30 + 5x$. Then, combine like terms: $8x + 5x - 12 - 30 = 13x - 42$. Simplify further: $13x - 42 = 7x - 27$.

Question 34: Solve the equation: $2(x + 3) - 4 = 12 - x$.

    a) $x = 4$

    b) $x = 5$

    c) $x = 6$

    d) $x = 7$

Answer: c) $x = 6$

Explanation: Begin by distributing 2 to the terms inside the parentheses: $2x + 6 - 4 = 12 - x$. Combine like terms: $2x + 2 = 12 - x$. Next, add x to both sides: $3x + 2 = 12$. Then, subtract 2 from both sides: $3x = 10$. Finally, divide both sides by 3 to solve for x: $x = 10/3 = 6$.

Question 35: What is the value of $(x^2 - 4)(x + 2)$ when $x = 3$?

    a) 35

    b) 42

    c) 45

    d) 54

Answer: b) 42

Explanation: Substitute $x = 3$ into the expression: $(3^2 - 4)(3 + 2) = (9 - 4)(5) = (5)(5) = 25$. Therefore, the value is 25.

Question 36: If $2(x + 5) = 4x - 6$, what is the value of x?

    a) -4

    b) -3

c) 4

d) 5

Answer: d) 5

Explanation: Begin by distributing 2 to the terms inside the parentheses: $2x + 10 = 4x - 6$. Next, subtract 2x from both sides: $10 = 2x - 6$. Then, add 6 to both sides: $16 = 2x$. Finally, divide both sides by 2 to solve for x: $x = 8/2 = 4$.

Question 37: Simplify the expression: $2x^3 + 4x^2 - 3x^3 + 5x^2$.

    a) $-x^3 + 9x^2$

    b) $-x^3 + x^2$

    c) $-x^3 + 5x^2$

    d) $-x^3 + 3x^2$

Answer: a) $-x^3 + 9x^2$

Explanation: Combine like terms by adding the coefficients of the same degree: $2x^3 - 3x^3 + 4x^2 + 5x^2 = -x^3 + 9x^2$.

Question 38: What is the solution to the equation $5(x - 2) = 3x + 6$?

    a) $x = 2/3$

    b) $x = 3/2$

    c) $x = 3$

    d) $x = 4$

Answer: b) $x = 3/2$

Explanation: Begin by distributing 5 to the terms inside the parentheses: $5x - 10 = 3x + 6$. Next, subtract 3x from both sides: $2x - 10 = 6$. Then, add 10 to both sides: $2x = 16$. Finally, divide both sides by 2 to solve for x: $x = 16/2 = 8/1 = 8 \div 2 = 4 \div 1 = 4/2 = 2$. Therefore, $x = 3/2$.

Question 39: Which of the following is equivalent to the expression $3x - 4(2x - 5)$?

    a) $-5x + 20$

    b) $-5x - 4$

    c) $-x + 20$

    d) $-x - 4$

Answer: c) $-x + 20$

Explanation: Distribute -4 to the terms inside the parentheses: 3x - 8x + 20. Combine like terms: -5x + 20.

Question 40: Solve the equation: 2(3x + 1) = 4 - 2(x - 3).

    a) x = -2

    b) x = -1

    c) x = 0

    d) x = 1

Answer: a) x = -2

Explanation: Begin by distributing 2 to the terms inside the parentheses: 6x + 2 = 4 - 2x + 6. Combine like terms: 6x + 2 = 10 - 2x. Next, subtract 2 from both sides: 6x = 8 - 2x. Then, add 2x to both sides: 8x = 8. Finally, divide both sides by 8 to solve for x: x = 8/8 = 1. Therefore, x = -2.

Question 41: Which of the following is the formula for calculating the area of a triangle?

    a) $A = l \times w$

    b) $A = \pi r^2$

    c) $A = \frac{1}{2}bh$

    d) $A = s^2$

Answer: c) $A = \frac{1}{2}bh$

Explanation: The formula for the area of a triangle is $A = \frac{1}{2}$ base $\times$ height.

Question 42: In a right triangle, if one of the acute angles is 30 degrees, what is the measure of the other acute angle?

    a) 30 degrees

    b) 45 degrees

    c) 60 degrees

    d) 90 degrees

Answer: c) 60 degrees

Explanation: In a right triangle, the sum of the measures of the acute angles is always 90 degrees. Since one of the acute angles is 30 degrees, the other acute angle must be 90 - 30 = 60 degrees.

Question 43: The sum of the interior angles of a triangle is:

    a) 90 degrees

b) 180 degrees

c) 270 degrees

d) 360 degrees

Answer: b) 180 degrees

Explanation: The sum of the interior angles of any triangle is always 180 degrees.

Question 44: A rectangular garden has a length of 10 meters and a width of 5 meters. What is its perimeter?

a) 10 meters

b) 20 meters

c) 30 meters

d) 40 meters

Answer: d) 40 meters

Explanation: The perimeter of a rectangle is calculated by adding all its side lengths. In this case, the perimeter is 2 × (length + width) = 2 × (10 + 5) = 2 × 15 = 30 meters.

Question 45: A circle has a diameter of 8 centimeters. What is its radius?

a) 4 centimeters

b) 8 centimeters

c) 16 centimeters

d) 32 centimeters

Answer: a) 4 centimeters

Explanation: The radius of a circle is half the length of its diameter. In this case, the radius is 8/2 = 4 centimeters.

Question 46: The formula for calculating the circumference of a circle is:

a) $C = \pi r^2$

b) $C = \pi d$

c) $C = 2\pi r$

d) $C = 2\pi d$

Answer: d) $C = 2\pi r$

Explanation: The formula for the circumference of a circle is $C = 2\pi r$, where r is the radius.

Question 47: Which of the following is the formula for finding the area of a circle?

   a) $A = l \times w$

   b) $A = \pi r^2$

   c) $A = 2\pi r$

   d) $A = 2\pi d$

Answer: b) $A = \pi r^2$

Explanation: The formula for the area of a circle is $A = \pi r^2$, where r is the radius.

Question 48: If a square has a side length of 6 centimeters, what is its area?

   a) 6 square centimeters

   b) 12 square centimeters

   c) 18 square centimeters

   d) 36 square centimeters

Answer: d) 36 square centimeters

Explanation: The area of a square is calculated by squaring its side length. In this case, the area is $6^2 = 36$ square centimeters.

Question 49: The sum of the interior angles of a quadrilateral is:

   a) 90 degrees

   b) 180 degrees

   c) 270 degrees

   d) 360 degrees

Answer: d) 360 degrees

Explanation: The sum of the interior angles of any quadrilateral is always 360 degrees.

Question 50: A right circular cylinder has a height of 10 centimeters and a radius of 3 centimeters. What is its volume?

   a) 30 cubic centimeters

   b) 60 cubic centimeters

   c) 90 cubic centimeters

   d) 180 cubic centimeters

Answer: c) 90 cubic centimeters

Explanation: The volume of a right circular cylinder is calculated using the formula $V = \pi r^2 h$. In this case, the volume is $\pi \times 3^2 \times 10 = 90$ cubic centimeters.

Question 51: Which of the following is a characteristic of a mammal?

a) Cold-blooded

b) Lays eggs

c) Has feathers

d) Produces milk

Answer: d) Produces milk

Explanation: Mammals are characterized by the ability to produce milk to nourish their young.

Question 52: What is the primary function of the circulatory system?

a) To transport oxygen and nutrients to cells

b) To produce hormones

c) To regulate body temperature

d) To remove waste from the body

Answer: a) To transport oxygen and nutrients to cells

Explanation: The circulatory system is responsible for the transportation of oxygen and nutrients to cells throughout the body.

Question 53: Which of the following is NOT an example of a renewable resource?

a) Solar energy

b) Wind energy

c) Natural gas

d) Hydroelectric power

Answer: c) Natural gas

Explanation: Natural gas is a non-renewable resource, meaning it is finite and cannot be replenished at the same rate it is consumed

Question 54: What is the smallest unit of matter?

a) Atom

b) Molecule

c) Cell

d) Electron

Answer: a) Atom

Explanation: An atom is the smallest unit of matter that retains the properties of an element.

Question 55: Which of the following is an example of a chemical change?

a) Freezing water into ice

b) Cutting a piece of paper in half

c) Burning wood

d) Melting chocolate

Answer: c) Burning wood

Explanation: Burning wood involves a chemical reaction that results in the release of heat, light, and the formation of new substances.

Question 56: What is the process by which plants convert sunlight into chemical energy?

a) Respiration

b) Photosynthesis

c) Transpiration

d) Fermentation

Answer: b) Photosynthesis

Explanation: Photosynthesis is the process by which plants use sunlight, water, and carbon dioxide to produce glucose and oxygen.

Question 57: Which of the following is an example of an exothermic reaction?

a) Melting ice

b) Boiling water

c) Rusting of iron

d) Photosynthesis

Answer: c) Rusting of iron

Explanation: The rusting of iron is an exothermic reaction that releases heat to the surroundings.

Question 58: Which of the following is an example of a conductor?

a) Rubber

b) Glass

c) Copper

d) Plastic

Answer: c) Copper

Explanation: Copper is a good conductor of electricity and heat.

Question 59: Which of the following is responsible for carrying genetic information in living organisms?

a) Proteins

b) Carbohydrates

c) Lipids

d) DNA

Answer: d) DNA

Explanation: DNA (deoxyribonucleic acid) carries the genetic information that determines the traits and characteristics of living organisms.

Question 60: What is the force that opposes the motion of an object through a fluid?

a) Gravity

b) Inertia

c) Friction

d) Acceleration

Answer: c) Friction

Explanation: When an object moves through a fluid (such as air or water), it experiences resistance or drag, which is a form of friction.

Question 61: Which of the following layers of the Earth is responsible for generating its magnetic field?

a) Crust

b) Mantle

c) Outer core

d) Inner core

Answer: c) Outer core

Explanation: The movement of molten iron and nickel in the outer core creates the Earth's magnetic field.

Question 62: What is the process by which rocks are broken down into smaller particles?

    a) Erosion

    b) Deposition

    c) Weathering

    d) Metamorphism

Answer: c) Weathering

Explanation: Weathering refers to the process of breaking down rocks into smaller particles through physical or chemical means.

Question 63: Which of the following is responsible for the formation of sedimentary rocks?

    a) Heat and pressure

    b) Cooling and solidification

    c) Weathering and erosion

    d) Deposition and compaction

Answer: d) Deposition and compaction

Explanation: Sedimentary rocks are formed when sediments, such as sand or mud, are deposited and undergo compaction over time.

Question 64: What is the primary source of energy for the water cycle?

a) Wind

b) Sun

c) Earth's core

d) Moon

Answer: b) Sun

Explanation: The Sun's energy drives the water cycle by providing heat that causes evaporation, leading to condensation, precipitation, and the circulation of water on Earth.

Question 65: What is the layer of the atmosphere closest to the Earth's surface?

    a) Stratosphere

    b) Thermosphere

    c) Mesosphere

d) Troposphere

Answer: d) Troposphere

Explanation: The troposphere is the lowest layer of the atmosphere, extending from the Earth's surface up to about 8-15 kilometers.

Question 66: Which of the following is a characteristic of a front in meteorology?

a) A sudden increase in atmospheric pressure

b) The boundary between two air masses with different properties

c) A circular storm system with low pressure at the center

d) The vertical movement of air currents

Answer: b) The boundary between two air masses with different properties

Explanation: A front is the boundary between two air masses with different characteristics, such as temperature, humidity, and air density.

Question 67: Which of the following is the unit used to measure astronomical distances?

a) Light-year

b) Astronomical unit

c) Parsec

d) Kilometer

Answer: a) Light-year

Explanation: A light-year is the distance that light travels in one year and is commonly used to measure astronomical distances.

Question 68: Which of the following is responsible for the formation of tides?

a) Gravitational pull of the Moon and the Sun

b) Earth's rotation on its axis

c) Solar flares and coronal mass ejections

d) Seismic activity in the ocean floor

Answer: a) Gravitational pull of the Moon and the Sun

Explanation: The gravitational pull of the Moon and the Sun on the Earth's oceans is responsible for the formation of tides.

Question 69: Which of the following automotive systems is responsible for igniting the fuel-air mixture in an engine?

    a) Ignition system

    b) Cooling system

    c) Fuel system

    d) Exhaust system

Answer: a) Ignition system

Explanation: The ignition system in an automobile is responsible for creating a spark to ignite the fuel-air mixture in the engine's combustion chamber.

Question 70: What is the purpose of the transmission system in a vehicle?

    a) To transfer power from the engine to the wheels

    b) To cool the engine

    c) To filter the air entering the engine

    d) To control the vehicle's suspension

Answer: a) To transfer power from the engine to the wheels

Explanation: The transmission system in a vehicle is responsible for transmitting power from the engine to the wheels and allowing the driver to change gears.

Conclusion

We have covered the most important topics and strategies that will help you pass the ASVAB test. We started by giving an overview of the ASVAB exam, its various versions and subtests. We also discussed how to interpret the ASVAB score and what to expect when you take the exam. We also discussed how ASVAB results determine military training and jobs.

We then delved deeper into the techniques that will help you pass the ASVAB test. We shared valuable tips on how to overcome test anxiety, and we also discussed strategies for answering multiple-choice questions. We stressed the importance of preparing for the test and creating a personalized study plan.

We then focused on communication, including topics like Word Knowledge and Paragraph Comprehension. We covered question formats, word meanings and vocabulary. We also provided tips and insights on reading comprehension strategies for slow readers.

Then we moved on to Math Skills, beginning with Mathematics Knowledge and Operations. We studied various mathematical concepts such as fractions, decimals and ratios. We also explored algebra, geometry and arithmetic logic. We gave you formulas and examples to solve math word problems.

We explored topics in the General Science section that were related to physical and life sciences. We studied biology, the human body, plant physiology and cells, genetics, chemistry and physics. To enhance your understanding, we introduced key scientific concepts and principles.

We then moved on to technical skills. We focused on auto information and shop information. We also covered mechanical comprehension, electronic information, and how to assemble objects. To prepare you for these subtests, we provided you with comprehensive knowledge about the topics.

We have included practice tests and practice questions in the book to ensure that you are ready. These exercises will allow you to put into practice the concepts and strategies that you have learned. They provide valuable experience. You can assess your performance by using the detailed explanations and answers.

You now have the knowledge and tools necessary to succeed on the ASVAB test. You can pursue your military career with confidence by following the study tips, practicing hard, and using the provided advice. Good luck on your journey.

References

Avital, M. (2005). Constructing a Computerized Adaptive Test for University Applicants with Disabilities. Applied Measurement in Education

Charles, P. (1981). Graphing and Linear Systems. Elsevier

Geoff, D. David, B. Ian, J. Effect of Breast Feeding on Intelligence in Children: Prospective Study, Sibling Pairs Analysis, and Meta Analysis. BMJ

Jaimie McMullen et al. (2022). Equity-Minded Community Involvement and Family Engagement Strategies for Health and Physical Educators. Journal of Physical Education

Jennifer, L. Power, R. (2003). ASVAB for Dummies

McMunn, R. H. (2013). ASVAB Mechanical Comprehension Test For Dummies. International Kindle Paperwhite

Mometrix Armed Forces Team. (2023). ASVAB Study Guide 2023-2024: ASVAB Test Prep Secrets. International Kindle Paperwhite

Rogers et al. (2022). Test Yourself in Essential Calculation Skills. Open University Press

The Princeton Review. (2017). The Princeton Review Math Workout for the New GRE. Princeton Review

Xiaonan Wang. (2022). Identification of Enclaves and Exclaves by Computation Based on Point-Set Topology. International Journal of Geographical Information Science

ASVAB. What To Expect When You Take the ASVAB. Department of Defense. www.officialasvab.com

Military.com. (2022). ASVAB Scores and Army Jobs. www.military.com

Made in the USA
Columbia, SC
18 July 2023